URBAN-RURAL INTEGRATION IN REGIONAL DEVELOPMENT:

A CASE STUDY OF SAURASHTRA, INDIA, 1800–1960

by

Howard Spodek
Temple University

THE UNIVERSITY OF CHICAGO
DEPARTMENT OF GEOGRAPHY
RESEARCH PAPER NO. 171

1976

Copyright 1976 by Howard Spodek
Published 1976 by The Department of Geography
The University of Chicago, Chicago, Illinois

Library of Congress Cataloging in Publication Data

Spodek, Howard, 1941–
    Urban-rural integration in regional development.
    (Research Paper–University of Chicago, Dept. of Geography; no. 171)
    Bibliography: p. 120
    1. Urbanization–Saurashtra, India (State). 2. Underdeveloped areas–Urbanization–Case studies. 3. Saurashtra, India (State)–Economic conditions. I. Title. II. Series: Chicago. University. Dept. of Geography. Research paper; no. 171.
H31.C514 no. 171 [HT147.I5] 910s [301.36′3′095475]
75-40468
ISBN 0-89065-078-0

Research Papers are available from:
The University of Chicago
Department of Geography
5828 S. University Avenue
Chicago, Illinois 60637
Price: $6.00 list; $5.00 series subscription

Dedication

To Susie, Joshua, and Sarah

ACKNOWLEDGEMENTS

I owe many thanks. My advisors, Professors Bernard S. Cohn and Brian J. L. Berry helped with encouragement and advice from the initial stages of planning the dissertation--on which this monograph is based--in Chicago in 1968 through to its completion in Philadelphia in 1972. I hope to repay Brian and Barney by being as generous toward my students as they were with me.

Librarians and archivists have made available materials and advice at the India Office Library in London, the National Archives of India in New Delhi, and District Record Offices in Saurashtra--Junagadh, Jamnagar, Bhavnagar, and particularly in Rajkot.

During our eight-month stay in Rajkot many people welcomed us with kindness and hospitality. Their help with research is reflected in this work; more personal attention can be repaid only by our warm appreciation. And we are pleased that our first child, Susannah Ruth, has Rajkot as her birthplace. Our sojourn there was eventful in personal as well as academic growth.

Whatever social concern and tenacity I bring to this work reflect in large measure my parents' legacy to me and I am grateful.

My family made major adjustments to my work schedule. I hope that they also gained from the travel and the experience. I thank them: my wife Marie and my children Susie, Joshua, and Sarah.

TABLE OF CONTENTS

ACKNOWLEDGEMENTS v

LIST OF TABLES ix

LIST OF ILLUSTRATIONS xi

INTRODUCTION 1

Chapter
I. SAURASHTRA ON THE EVE OF BRITISH RULE 4

    Cities as Centers of Trade and Manufacture
    Actors and Interests in the Urban Areas
    The Interactions of Urban Groups
    Town and Village
    Alternative City-State Politics
    Conclusions

II. THE EFFECTS OF BRITISH POLICY ON THE URBAN SYSTEM OF
SAURASHTRA 34

    Princely Politics
    Effects of British and Princely Politics
    Summary and Conclusions

III. SAURASHTRA SINCE INDEPENDENCE: THE ACHIEVEMENT OF
AN URBAN-RURAL REGIONAL BALANCE 60

    Introduction
    The Infrastructure
    Land Policy
    Urban Development
    Conclusions

IV. CASE STUDY OF A SINGLE CITY-STATE 92

    Economic Development of Rajkot
    Industrialization and Trade
    Rajkot and Urban-Rural Integration

V. HISTORY AND URBAN THEORY 105

    Conclusions

APPENDIX	117
BIBLIOGRAPHY	120

LIST OF TABLES

1. Percentage of the Population of the Saurashtra Peninsula Living in Urban Areas, with All-India Comparisons, 1901-1971 2
2. Urban Population of Saurashtra, 1872-1941 41
3. Municipal Expenditures 46
4. Total Sea Trade in Rupee Value 47
5. Miles of Railway and Total Rail Freight Tonnage 48
6. Number of Industries in Cities 50
7. Percentage of Revenue from Land, Customs, and Railroads 51
8. Emigration from Kathiawad--1911 55
9. Salt Exports 63
10. Road Mileage 65
11. Road Travel Facilities--1948-57 66
12. Post Offices 66
13. Schools 67
14. Distribution of Schools in Village Areas--1961 67
15. Banking Offices 68
16. Electrical Generation 68
17. Literacy 69
18. Newspapers Published 69
19. Printing Presses 70
20. Cargo Handled 71
21. Joint Stock Companies 72
22. Factories 73

23.	Changes in the Number of Holdings of All the Categories of Cultivators between 1947-48 and 1954-55	75
24.	Incidence of Taxation per Capita	80
25.	Panchayats, 1950-51 and 1960-61: Number, Income, Expenditure	81
26.	Area under Various Crops (Acres)--Saurashtra	82
27.	Area under Various Crops (Acres)--Bhavnagar	82
28.	Area under Various Crops (Acres)--Junagadh	83
29.	Area under Various Crops (Acres)--Nawanagar	83
30.	Exports of Groundnut Products	87
31.	Distribution of Factories by Major Industries--1961	87
32.	Members of Kunbi (Patel, Patidar) Caste Enrolled at Alfred High School, Rajkot	88
33.	Sex Ratios	90
34.	Size Ranking of Saurashtra Cities over 20,000, 1872-1961	93
35.	Trade of Rajkot State	96
36.	Organized Industries in Rajkot State	97
37.	Acreage under Groundnut--Three State Areas	102
38.	Cotton Cultivation in Rajkot	103
39.	Amount Borrowed from Village Banks	104

LIST OF ILLUSTRATIONS

Map
1. Saurashtra . . . . Frontispiece

2. First and Second Class Roads Constructed by the British between 1865 and 1880 42

3. Railways in Saurashtra, 1880-1900: The First Two Decades 49

Graph
1. Distribution of Land Holdings, 1947-48 and 1954-55 76

2. Distribution of Land in Various States of India, 1953-54 78

3. Size Ranking of Saurashtra Cities, 1872-1961 95

Figure
1. A Hierarchy of Market Centers and Market Areas 108

2. A Model of the Process Generating Increased Interaction between Large Cities 111

3. The Circular and Cumulative Feedback Process of Urban-Size Growth for the Individual American Mercantile City, 1790-1840 112

INTRODUCTION

This study is a composite of three themes, inter-related and moving from concrete and particular to theoretical and general. At its core is a detailed case study of the inter-relationships of cities and the countryside surrounding them in Saurashtra, India from 1800 to 1960. It examines the effects of a variety of government structures and policies on these inter-relationships. This core section is based on direct personal research in archival and newspaper collections in London, Saurashtra, and New Delhi as well as on interviews among people affected.

The Saurashtra peninsula of western India presents a peculiarly appropriate arena for studying the impact of diverse political structures on urban-rural interaction. Saurashtra is, for India, a highly urbanized area with urban roots stretching back for millennia. At least as far back as the proto-historical Harappan era of 1500 B.C. cities have been known in Saurashtra.[1] In modern times, beginning with the census of 1872, Saurashtra has registered a percentage of urban population almost double that of India generally. In 1961, with only about 1 per cent of the nation's population and land area, Saurashtra had 3 per cent of all cities over 20,000 population and 4 per cent of all those over 100,000 (see Table 1).

Also Saurashtra has experienced a variety of political structures during the 160-year period under study; in light of my concern for political structures as a key to determining the role of the city in the economic life of the region, this provides a framework for comparison. Three major political forms can be distinguished chronologically. Before the British proclaimed their paramountcy over Saurashtra in 1820, and in many respects even up to 1867 when they asserted themselves actively, Saurashtra was governed by independent princes who carved the peninsula into numerous petty "rajadoms." Many of these "rajadoms," including all of the larger ones, took pains to build capital towns, a fact

---

[1] Sir Mortimer Wheeler, <u>Civilizations of the Indus Valley and Beyond</u> (New York: McGraw-Hill Book Co., 1966).

TABLE 1

PERCENTAGE OF THE POPULATION OF THE SAURASHTRA
PENINSULA LIVING IN URBAN AREAS, WITH
ALL-INDIA COMPARISON, 1901-1971

Year	Total	All-India	In Towns over 20,000	All-India in Towns over 20,000	In Cities over 100,000	All-India in Cities over 100,000
1901	26	11	10	5.5	..	2
1911	23	10	9	5	..	2
1921	24	11	9.5	6	..	3
1931	25	12	13.5	7	..	3
1941	28	14	16	9	3	5
1951	31	17	20	12	8	7
1961	31	18	22	14	9	9
1971	31	20	N.A.	16	10	10

Sources: Census of India for appropriate years.

which largely accounts for the pre-British urbanization of the region. Rule by rajas was highly localized, limited to the area they could control militarily. Chapter I discusses this pre-British era.

British rule brought an overarching political structure which incorporated these rajadoms and introduced new cultural patterns as well as the opportunity for increased contact with the world outside Saurashtra. Nevertheless, the British chose to rule Saurashtra indirectly; they kept alive the system of princely rule and allowed a marked degree of internal autonomy to these rulers. The colonial period is considered in Chapter II.

The third period began in 1947 with the advent of independence and, in the next year, the dissolution of the princely states and their merger into one political union tied firmly into the national structure. Thus Saurashtra offers wide scope for comparative study of urban development, urban-hinterland influences, intra-regional activities, and ties between the region as a whole and other regions outside. Chapter III discusses post-Independence conditions.

Chapter IV focusses on a single city-state, Rajkot, a small political jurisdiction centered on an administrative capital. Throughout the period 1800-1960 Rajkot's geographic centrality led to its choice by outside governments--Maratha, British, and the independent State of Saurashtra--as the location for their regional capital as well. Again, political decisions proved crucial to urban development, regional patterns of growth, and urban-rural interaction.

Following the detailed case study of Saurashtra and Rajkot, a second theme is developed relating these historical insights to current theories in regional and urban planning. These theories, discussed at length in Chapter V, argue that regional economic development is best stimulated through the judicious creation of new urban nodes and the enrichment of older ones. Based largely on an application of the geographers' central place theory, they urge the creation of numbers of urban nodes, from small market towns nestled in rural areas where heretofore farmers have not had access to markets, up through middling sized towns to provide larger regions with economic and administrative facilities and keep them in touch with still larger centers of innovation and creativity. Additional theoretical views support the significance of cities in economic development: Cities serve as centers of innovation and dissemination of innovation because they are the hubs of communication and transportation networks, and cities' specialization of labor and efficiency of production engender economic growth. These theories complement the policies for urban-centered developmental policies in lagging areas. They propose assistance to existing cities and the creation of new ones to stimulate creativity, productivity, market orientation, and national integration.

Saurashtra's experience suggests areas of potential difficulty in applying urban-based planning models. In Saurashtra political variables constantly intervene between planning theory and reality. Cities are not independent variables, but rather function primarily within the parameters set by government policies. Government may be too weak to formulate and execute developmental policies, or to base them on urban growth schemes in countries which are heavily rural. Indeed, economic growth may not even be their prime consideration. To the extent that national governments may be weak or ambivalent, as they often are in technologically underdeveloped countries, the experience of Saurashtra may be relevant. Developmental policies based on urbanization may be formulated only in the context of political realities. Conclusions drawn by comparing contemporary planning theories with historical case study are presented in the final chapter.

CHAPTER I

SAURASHTRA ON THE EVE OF BRITISH RULE

> The coincidence and conflict of interest between king and merchant emerges as an important feature of the traditional city which should, I think, provide a background for the discussion of modern capitalistic activity in the modern city.
> --David Pocock, "Sociologies--Urban and Rural"

On the eve of the British arrival in Saurashtra at the end of the eighteenth century, the cities of the peninsula served as nodes of two systems of power. One was based on landholding, headed in most places by a chief of a Rajput-caste lineage who made the city his capital and armed fortress. Since control of land was an important goal for the Rajputs, competition among potential landholders, both between neighboring rulers and between ruling chiefs and underlings within the lineage group, introduced a constant tension into the political-military balance of the region. Sometimes war resulted; more often competitive rulers and aspirants peacefully negotiated settlements of their claims to shares in the land and its produce. Settlements were usually at the expense of the weakest and usually from the produce of the land of the least protected peasant. But the urban based ruler could not simply exploit the peasantry at will. The sparse population of Saurashtra induced competition for peasants to farm the land; the peasants could emigrate or threaten to emigrate in the face of excessive exploitation. The ruler had to be responsive to peasant demands.

Few towns supported activities beyond this administrative function. Most served only as centers of local rule presided over by a member of the lineage elite of the dominant Rajput group. They provided a revenue collecting, administrative link between the higher level government and the local population. But they fostered little trade and manufacture; they did little to welcome members of these professions. They remained "rurban."

Cities with rulers of greater power and imagination attracted a second group of elites as well; a truly urban elite based not on control of land but on control of capital, expertise, and personal connections. Its participants were highly mobile, seeking jobs where the conditions were best, and employing the threat of emigration as a major lever of power. Through caste and occupational affiliation these elites had ties which spread widely across geographical space. Each group had its own sphere of interests--rulers, their relatives, merchants, moneylenders, professionals, artisans, and soldiers. To some extent these interests were separable, as reflected in the division of the city into <u>mohullas</u>, small, distinct neighborhoods, where groups lived separately from one another. But to some extent interests converged, so that rulers often had to accept compromise or face the possible loss of valuable subjects. The vast number of competitive capitals in a small area gave opportunity to those urbanites willing to migrate, and the relatively sparsely settled land offered similar prospects to farmers. The city, and even the countryside, were not so much "home" as a base of operation.

The city was the cockpit in which all of these diverse groups worked, sometimes for their mutual benefit, sometimes in opposition to one another. Struggles for power declined when one authority inside or outside the peninsula could assert and maintain paramount power. No single pattern characterized all the cities of the peninsula at any one time nor any one major city through extended periods. Change and shifting balance were the normal pattern. Around 1800, no strong, peninsula-wide authority existed; power was contested in both the territorial and mobile systems. The multiplicity of small competitive city-states in the region promoted both military instability and the opportunity for great personal mobility--physical, occupational, social, and economic.

Toward the end of the eighteenth century, Saurashtra was largely independent of outside authority. Internally it was carved into a multitude of feuding, discrete city-states each of which was controlled by a local warrior-ruler. In 1787, the (evidently first) British emissary to Saurashtra saw the peninsula studded with towns which served primarily as military fortresses. Dr. Hove, a Polish scientist dispatched by the British to assess the prospects of cotton purchases from Saurashtra,* described Limbdi, a major city in the midst of

---

*Professor Holden Furber informed me that Hove's real, but secret, mission was to determine whether Indian varieties of cotton could be grown in the West Indies. The secret orders as well as the public ones, Furber noted, are preserved in the Public Record Office.

the cotton producing tracts of eastern Kathiawad:

> The town is above five miles in circumference, built in a square, and surrounded by a thick wall of stone. The wall has 16 towers, two at each of the two entrances of the town; and the rest are divided, without any regularity, at all sides of the wall. Eight of these towers mounted eight iron guns, which were cast in their place. On the inside of the wall is a rampart for the musquetry, not unlike that of Broach and Cambay, with divisions, and a breastwork, by which they are well-protected from their assailants. The ditch which encompasses the walls is about 40 feet broad, and 30 deep, and is usually filled in the rainy season by the communication it has with the river. The raja maintains a constant army of 13,000 men, the greatest part horsemen, and the rest matchlockers, but they are all undisciplined. . . .[1]

Twenty years later a more comprehensive report by [then] Colonel Alexander Walker, the British Resident at the court of Baroda, identified 162 places as fortified; 97 had towers. In 1849 another survey showed 921 towns containing fortified residences of the darbar, of which 79 were towns having walls with bastions and 16 were towns having other fortified places within them.[2]

The defenses were necessary; undefended areas were threatened by many groups. One threat to local rulers was the attempt of outside powers to rule Saurashtra or at least to extract tribute. As a peripheral region of the heartland of the Indian plains, Saurashtra fell into the orbit of an all-India power only when the central power had enormous strength, as did the Emperor Akbar whose armies conquered Saurashtra in 1591-92, for example.[3] Otherwise it might fall to the nearest external power, for example the Sultan of Gujarat before Akbar. At the time of the arrival of the British, the neighboring external power, the Marathas, were powerful enough to collect tribute from the province but not to establish an administrative apparatus for the peninsula.[4]

---

[1] Dr. Hove, Tours for Scientific and Economical Research Made in Guzerat, Kattiawar, and the Conkuns in 1787-88: Selections from the Records of the Bombay Government, Vol. XVI, New Series (Bombay: Government Central Press, 1855), p. 77.

[2] Selections from the Records of the Bombay Government, XXXIX (New Series, 1856), 339. This volume was reprinted by the government of Bombay in 1894, but renumbered as XXXVII. This is confusing because in the same reprint series that volume XXXVII was renumbered XXXIX. Thus the two volumes were interchanged. Unless otherwise noted, the 1856 volumes and page numbering are used.

[3] S. C. Misra points out that the continuing fragmentation of Rajput states in Saurashtra during the seventeenth century suggests that the Mughal authority was "superficial." Gujarat State Gazetteers, Rajkot District (Ahmedabad), 1965, p. 37.

[4] For a regional approach to Indian geo-politics, see Bernard S. Cohn, "Regions Subjective and Objective: Their Relation to the Study of Modern Indian

Neighboring rulers within the peninsula itself posed another threat; they often attempted to expand their territory at the expense of weaker neighbors. The states' political and tributary inter-relationships were complex, but they appeared to rest on military strength; if one grew more powerful it might seize land from its neighbor, or at least successfully demand a share of tribute from the neighbor's land revenue. The head of the British military and administrative mission to pacify Saurashtra and establish treaty and tribute relationships with the local states, summarized the situation around the Saurashtrian shore of the Gulf of Cambay in 1804:

> The district of Gogo is a collection of independent states, the chief of which is the Raja of Bhavnagar. This Raja and the rest are grasias [holders of inalienable rights in the land. They also had legal jurisdiction over their tenants] but they owe no obedience to each other, unless what they may contract by voluntary engagements. They generally reside in places of difficult access, and some of them have built extensive stone fortifications which are, however, indifferently provided with cannon, as well as deficient in other means of defense. The Raja of Bhavnagar entertains in his service about 7,000 infantry and 5 or 600 cavalry. Most, if not the whole, of these states pay Moolukgeeree contributions to the Marathas, and to the Nawab of Junagadh, besides the Jumma or taxes which they paid to the Peshwa, and which have been ceded to the [British East India] Company.[5]

The greatest number of political and tribute relationships and their shifting balances encouraged the threat of attack and reinforced the need for defense: "It may easily be conceived that where there are so many claims the authority of this country is divided in a very singular manner; and it would not, perhaps,

---

History and Society," in Regions and Regionalism in South Asian Studies: An Exploratory Study, ed. Robert I. Crane, pp. 5-37.

Histories of Saurashtra are few. H. Wilberforce-Bell, The History of Kathiawar (London: William Heinemann, 1916) is the only one in English. The standard gazetteer, Kathiawar: Gazetteer of the Bombay Presidency, Vol. VIII (Bombay: Government Central Press, 1884) provides much useful information. Another useful set of books are the mini-gazetteers of five of the major states of the peninsula prepared by J. W. Watson as background for the overall gazetteer of Kathiawar. They are entitled Statistical Account of Bhavnagar, Statistical Account of Junagadh, Statistical Account of Porbandar, Statistical Account of Nawanagar, and Statistical Account of Dhrangadhra. All the volumes were published in Bombay between 1883 and 1885. An especially useful historical perspective is Ranchodji Amarji's Tarikh-i-Sorath (Bombay: Education Society's Press, [1882]), an English translation from the Persian. Ranchodji's father was one of the greatest diwans of Saurashtra's largest state, Junagadh, from the 1760's until his assassination in 1784.

The most recent and comprehensive history in Gujarati is Shambhuprasad Harprasad Desai, Saurashtrano Itihaas (Junagadh: Sorath Shikshan Ane Sanskruti Sangh, 1968). It lacks, however, an integrated approach, tending rather to be a catalogue of events state-by-state.

[5] Selections, XXXIX, 21.

be an easy matter to point out the paramount power."[6] The lack of an accepted balance of power in the political-military sphere encouraged the possibility of attacks. Conflicting jurisdictions, autocratic legal systems, weak governments, and ill-defined borders all discouraged peaceful settlement.

Banditry was a third threat against which fortification was required. Dr. Hove reported in 1787:

> The strength of the kolis [a farming caste] and grassias seems to have been still occasionally too great to allow of Guzerat [and Saurashtra as well] being considered a settled country; cattle-liftings, and such like irregularities were looked on as daily occurrences; and a stranger could not stir, to travel from one place to another, without a guard of some twenty or thirty kolis or horsemen.[7]

Among the looters were the bahaarvatiyos, literally outlaws, who were guerrilla warriors in revolt against their rulers. Such groups rose periodically in every state to attack rulers who were considered unjust. Because the bahaarvatiyos limited their attacks only to the property of the ruler and did not disturb the poor, they had the warm regard in Saurashtra which the Robin Hood legend conveys to Westerners. Indeed, the account of the bahaarvatiyos given by the British official Kincaid suggests that their cause was usually just.[8] Jhaverchand Meghani's four volumes of tales about them enshrine them as folk heroes.[9]

Sometimes bahaarvatiyos were ex-rulers reduced to nomads by the loss of their former citadel. For example, when the territory of Rajkot-Sardhar fell to Masumkhan, the deputy Faujdar of Sorath, the former ruler, Thakor Meheramanji, now without a capital, went into bahaarvatun. Finally, twelve years later, in 1732, his seven sons succeeded in reconquering Rajkot, and, a few years later, retook Sardhar as well.

In other cases, going into outlawry was the first step taken by a junior member of a lineage in an attempt to wrest land for himself from his lineage chief. Thus, for example, the Gondal State was formed as a breakaway splinter from Jamnagar in 1560. The Rajkot State was established by Thakor Shri Vibhaji, also a scion of the Jamnagar House in 1617. In these two cases, as in others, the areas carved out were on the borders between feuding larger rulers.

---

[6] Ibid.

[7] Hove, p. ix.

[8] C. A. Kincaid, The Outlaws of Kathiawar and Other Studies (Bombay: Times Press, 1905).

[9] Jhaverchand Meghani, Sorathi Bahaarvatiyaa (4 vols.; Ahmedabad: Gurjar Grantharatna Kaaryaalaya, 1929).

Such areas were easier to seize because they were less firmly controlled and since they were disputed they offered possibilities of alliance with at least one of the larger parties to the dispute. They symbolized the lack of political integration in the peninsula. Whatever the justice of banditry, so arbitrary a system of rule and protest left the countryside unsettled.[10]

Not that warfare was continuous. Indeed, local rulers sometimes assisted one another--for a price. The records of Gondal State and Junagadh State both report that around 1760, when Junagadh was particularly weak, Gondal provided armies to fight off neighboring attackers and with money and troops to pay and subdue Junagadh's own rebellious forces. In exchange, Gondal was granted control over the parganas of Upleta and was assured of the continuance of a weak, non-threatening ruler continuing as chief of her largest neighbor.[11]

Perhaps the political condition in Saurashtra on the eve of the British arrival could be described as "cold warfare" with intermittent outbreaks of "hot war." As Major Walker observed in 1804, negotiation, agreement, and a balance of power short of war marked the region generally:

> It might be expected that such intermingled interests would produce internal causes of disputes and quarrels, but such is the discretion, or habits of the people, that the respective shares of the parties have been usually collected and quietly accounted for, without any contention.[12]

---

[10] Shri Yaduvansh Prakaash, p. 41. No publication data found. The book is a bardic history of the Jadeja Rajputs who dominate northwestern Saurashtra. It was compiled in the mid-1930's and published in either Rajkot or Jamnagar. My copy lacked that information.

The bahaarvatiyos of Saurashtra correspond to the "Social Bandits" studied in Europe by E. J. Hobsbawm in Primitive Rebels (New York: W. W. Norton and Co., Inc., 1965), pp. 13-29. Cf. Hobsbawm's description of the conditions which give rise to social banditry: "It is rural, not urban. The peasant societies in which it occurs know rich and poor, powerful and weak, rulers and ruled, but remain profoundly and tenaciously traditional, and pre-capitalist in structure. . . . Moreover, even in backward and traditional bandit societies, the social brigand appears only before the poor have reached political consciousness or acquired more effective methods of social agitation. The bandit is a pre-political phenomenon and his strength is in inverse proportion to that of organized agrarian revolutionism and Socialism or Communism. . . .

"In such societies banditry is endemic. But it seems that Robin-Hoodism is most likely to become a major phenomenon when their traditional equilibrium is upset; during and after periods of abnormal hardship, such as famines and wars, or at the moments when the jaws of the dynamic modern world seize the statis communities in order to destroy and transform them." (Pp. 23-24.)

[11] Shri Yaduvansh Prakaash, Section 2, pp. 104-8. See also Watson, Statistical Account of Junagadh, p. 35.

[12] Selections, XXXIX, 21.

Unfortunately, these agreements were reached by trading off the income of the cultivators to buy protection: "While such arrangement, however, exists, . . . the prosperity of the country must be sacrificed."[13]

The agreements, once reached, did not always hold, as was made clear to the Company servants. In May 1806, girasias, apparently inspired by the Maharaja of Bhavnagar, attacked and burned the settlement which the Company was fostering near Dholera. Walker then decided that until the vast number of petty independent states disappeared and one system of justice and rule was established, property and order could not be secured.[14] Thus at the eve of the British assumption of paramountcy, princes of territory stood in open opposition to one another in an uncertain relationship always hovering near the brink of war, except when one power or another could assert decisive military supremacy.

No wonder, then, that fortified towns studded the Saurashtra landscape. And no surprise that they contrasted sharply with the open, exposed countryside. Alexander Kinloch Forbes, the folklorist of Gujarat, recounts the success of the fortress-city in withstanding these demands for tribute:

> A Moolukgeeree army seldom possessed power sufficient to subjugate a country, or to reduce its fortresses, which were sure to be defended with obstinacy; it carried on its operations, therefore, against the open towns and villages, selecting the season of harvest for its period of action, with the view not only of compelling more speedy acquiescence of the chieftain, but also of securing the more ready means of subsistence for the troops.[15]

The oft-noted gap between the ruling city and its hinterland resulted from a political system which attempted to assign much of the wealth of the countryside to fortress-based, often urban-based, rulers.

The uneasy balance of force which resulted had important consequences for the urban structure, particularly for the sizable proportion of it which served as political capitals. Town rulers saw each other as enemies and contestants for a relatively inelastic good, control over land revenues and tenants, and they often opposed one another. Their fortresses and their armies were

---

[13] Ibid.

[14] Pamela Nightingale, Trade and Empire in Western India, 1784-1806 (Cambridge: Cambridge University Press, 1970), p. 231. Cf. also J. H. Gense and D. Banaji, eds., The Gaikwads of Baroda: English Documents (10 vols.; Bombay: D. B. Taraporevala Sons and Co., n.d.), VII, 530.

[15] Alexander Kinloch Forbes, Ras Mala (London: Richardson and Company, 1878), pp. 394-95, and cf. also James Tod, Travels in Western India (London: W. H. Allen and Co., 1839), p. 305.

symbols of the multiple geo-political divisions in the peninsula.

The tiny Saurashtra principalities had few of the attributes that led in Europe to the creation of nation-states. They were so near to one another geographically that they did not differ substantially in language, customs, or in caste composition. Loyalty to the "state" was minimal, consisting perhaps of loyalty to ancestral land and to the current ruler. In addition, as Fox points out, adherence to restrictive marriage and kinship patterns may have been the factor which prevented Rajput rulers from establishing a bureaucratic-territorial state rather than just a system of control over land rights.[16] Indeed, in Saurashtra even marriage ties did not appear to ameliorate the threat of war. Major Walker noted that marriages among the families of Rajput chieftains were common and produced temporary amnesties and peace, but they did not produce "lasting and cordial interests."[17]

Cities as Centers of Trade and Manufacture

The prime function of cities in Saurashtra was to provide a fortress for defense, but many of the cities housed additional functions, notably trade. International trade had long been important to the peninsula which has a 600-mile coastline, one-fifth of India's total. In 1580, a high point, government revenue from the trade was estimated at Rs. 1,400,000. If this figure represents a 5 per cent tariff, as was customary, then total sea trade equalled Rs. 28,000,000.[18] Subsequent political decentralization and the concentration of coastal trade at the mainland ports of Surat and, later, Bombay, curtailed but did not terminate the importance of Saurashtra's ports.

Bhavnagar was the chief example of a city which successfully united strong military rule with trade concessions. Bhavsinhji, the chief of the Gohel Rajputs, successfully defended his capital, Sihor, from Maratha attack, but realized that the city was not an easy one to defend. He chose the site of Bhavnagar for a new capital and began construction there in 1723. His success in founding a new capital and commercial emporium underlined the potentials for

---

[16] Richard G. Fox, Kin, Clan, Raja, and Rule: State-Hinterland Relations in Pre-Industrial India (Berkeley: University of California Press, 1971).

[17] Selections, XXXIX, 109.

[18] Michael Naylor Pearson, "Commerce and Compulsion: Gujarati Merchants and the Portuguese System in Western India, 1500-1600" (unpublished Ph.D. dissertation, University of Michigan, 1971), p. 40.

the combination of the two functions within a city. Trade usually required strong government support. Compare the converse experience of Bhavnagar's neighbors: As Bhavnagar was gaining, "The trade of Gogo and Cambay had proportionately decayed as those ports were deprived of protection and unsupported any longer by the lucrative communication with [the provincial capital] Ahmedabad."[19] Compare also the treatment of piracy in Bhavnagar with that in other major states of Saurashtra: The Nawab of Junagadh, finding his forces too meager to protect sea trade, closed the most infested ports.[20] Bhavnagar, facing a similar threat, entered into a cooperative alliance with the British to root out the pirates.[21] Nawanagar made an alliance with the pirates![22] In familiar terms, the flag and trade had to move together.

Opinions conflict on the degree to which this combination of military and economic power was attained in other cities. On the one hand, Captain George LeGrand Jacob, Political Agent of the British in Saurashtra, 1839-43, before the British began the active administration, wrote that "The traditional policy of the state was to maintain inaccessibility. Forests, difficult passes, vile roads, thick jungles, were the bulwarks not only of the capital but of most of its towns and villages."[23] On the other hand, the Tarikh-i-Sorath, a history of Saurashtra written in the opening years of the nineteenth century by the son of a major diwan, noted that Nawanagar State held three ports and worked an iron mine, presumably for the manufacture of weapons.[24] A few years later, Major Walker noted that Nawanagar also manufactured coarse cotton cloth both for domestic use and for export. He, too, noted that Nawanagar (and Morvi and Kutch as well) were celebrated for excellence in steel manufacture. The seaports and principal towns, Walker wrote, possessed some men of wealth and property.[25]

---

[19] Forbes, Ras Mala, p. 418.

[20] Selections, XXXIX, 191.

[21] Ibid., pp. 149-56.

[22] Nightingale, p. 206.

[23] George LeGrand Jacob, Western India: Before and during the Mutinies (London: Henry S. King and Co., 1872), p. 121.

[24] Ranchodji Amarji, pp. 298-99.

[25] Selections, XXXIX, 273.

Junagadh, though it had been founded on a remote, forested hill which well fit Jacob's description of inaccessibility, nevertheless had a great variety of artisans producing textiles, turbans, headdresses, dhotis, and petticoats. They manufactured silk stuffs, like those of Ahmedabad and Surat, called mashru, atlas, and panchpata. Painters, dyers, workers in shells, engravers, embroiderers, and tailors had their workshops as did perfumers. Bazaars had an abundance of produce and Bohoras, Khatris, and Bhatia castemen carried on trade.[26] Ideally, then, trade and rule reinforced one another. Defensive and military headquarters seemed, however, the sine qua non for a flourishing town; without adequate military protection, trade would wither.[27]

Most studies of the structure of government in Mughal and post-Mughal India have stressed the land-holding elements in the administration. Even the titles of some of the most significant works reflect this concern, e.g., Irfan Habib's Agrarian System of Mughal India[28] and W. H. Moreland's Agrarian System of Moslem India.[29] Sir Jadunath Sarkar's pathbreaking Mughal Administration[30] follows a similar analysis. Our discussion so far has pointed out that land control and military supremacy were necessary to local administration; but other important activities, notably trade, also marked Saurashtrian public life especially urban life. We now turn to an investigation of the public life, actors, and groups in urban Saurashtra.

Actors and Interests in the Urban Areas

Hints of the internal organization of the towns by neighborhoods appear in the Ras Mala:

---

[26] Ranchodji Amarji, p. 244.

[27] Cf. Gideon Sjoberg, "The Rise and Fall of Cities: A Theoretical Perspective," International Journal of Comparative Sociology, IV, No. 2 (September 1963), 107-20. Sometimes in the absence of adequate governmental protection, however, merchants hired their own guards. James Tod, Annals and Antiquities of Rajasthan (2 vols.; London: George Routledge and Sons, 1914), I, 379, and Hove, Tours.

[28] Irfan Habib, The Agrarian System of Mughal India, 1556-1707 (Bombay: Asia Publishing House, 1963).

[29] W. H. Moreland, The Agrarian System of Moslem India (Delhi: Oriental Books Reprint Corporation, 1968).

[30] Sir Jadunath Sarkar, Mughal Administration (Calcutta: M. S. Sarkar and Sons, Ltd., 1935).

The towns are usually surrounded by a wall, and divided internally into mohullas, or wards, each of which contains many houses, but has only one public gateway, and constitutes a species of inner castle. The only public buildings, with the exception of government offices, are those which are devoted to religious purposes--mosques, temples, serais, Jain convents.[31]

The paucity of public buildings for the entire civic community suggests that small, more localized groups were crucial to the life of the city.[32] This evident existence of sub-groups within the population by occupation and by residence prompts the question: What were the leading groups in urban life?

First, we must note the rulers who held power through controlling land by military superiority. Their statuses rose and fell in proportion to their ability to maintain military power. Chieftains rose in power by gaining cessions of land and land rights from weaker rulers in exchange for protection.[33] This, of course, was the main British complaint against them; rather than trying to improve the land, they seemed constantly searching for weakness in a neighbor which would enable attack for more land.[34] Rulers over large holdings built and lived in their capital cities but drew their power largely from control of land held by military power.

In times of uncertain rule, especially, the kshatriya-caste warrior-rulers, represented locally by the Rajputs who controlled most of the land of Saurashtra, had reason to live in accordance with the "ideal" code of active militance. According to Max Weber in his Religion of India, the original concern of the kshatriya was the welfare of his subjects, whom he protected politically and

---

[31] Forbes, Ras Mala, p. 551.

[32] The internal division of cities into separate distinct neighborhoods based on kinship or on occupation has been noted repeatedly in a wide variety of preindustrial cities. Cf. Bernard S. Cohn, "Political Systems in Eighteenth Century India: The Banaras Region," Journal of the American Oriental Society, LXXXII, No. 3 (1962), 312-20; Ira M. Lapidus, Muslim Cities in the Later Middle Ages (Cambridge: Harvard University Press, 1967); Ira M. Lapidus, "Muslim Cities and Islamic Societies," in Middle Eastern Cities, ed. Lapidus (Berkeley: University of California Press, 1969); Wolfram Eberhard, Settlement and Social Change in Asia (Hong Kong: Hong Kong University Press, 1967), pp. 43-64; Gideon Sjoberg, The Preindustrial City (New York: Free Press, 1960), pp. 91-103.

[33] Selections, section on Jhalavad. Cf. also Gense and Banaji, VII, 496.

[34] Cf. Walter C. Neale, "Land Is to Rule," in Land Control and Social Structure in Indian History, ed. Robert Frykenberg, pp. 3-16, for a general statement on this attitude of rulers.

militarily.[35] An early view was: "That king is good whose subjects are prosperous and experience no famine."[36] Later, however, the kshatriya developed a vocational duty far more narrow:

> Warfare is the dharma of the kshatriya in classical and medieval sources. Except for the intermissions brought by the universal monarchies, war was ever-present in India as between the ancient city-states . . . That a king should ever fail to consider the subjugation of his neighbors by force or fraud remained inconceivable to secular and religious Hindu literature.[37]

The bias of the local historical accounts preserved by bards of the Barot, Bhat, and Charan castes further accentuated the military and aggressive concerns of their Rajput patrons. "The main themes of bardic ballads are battles, warriors, and kings."[38] Accounts of different ruling houses do differ, however, in emphasis. The accounts of the Jadeja clans of Jamnagar focus much more on military exploits than do those of Bhavnagar, for example. The result is that we know of a great regard for commerce in Bhavnagar, from Indian-authored histories as well as from British writings.[39] At times when one ruler was powerful enough to establish sway over substantial territory and be free from worry from his immediate neighbors, he might well adopt a less militant posture. But in cases such as Saurashtra in the late eighteenth century, the Weberian "ideal" seems to accord very well with practice.

Apart from the rulers, yet within the ruling family, lay another source of potential power: the harem. In recounting the intrigues of the courts of Kathiawad, the Tarikh-i-Sorath included an alliance between the wife of the Jam Saheb of Nawanagar and her father the Rao of Halavad, against her husband.[40] Since both Rajput rulers and Muslim rulers usually took several wives, the struggles

---

[35] Max Weber, The Religion of India (Glencoe, Ill.: Free Press of Glencoe, 1958).

[36] Ibid., p. 64.

[37] Ibid., and p. 146.

[38] A. M. Shah and R. C. Shroff, "The Vahivanca Barots of Gujarat," in Traditional India: Structure and Change, ed. Milton Singer (Philadelphia: American Folklore Society, 1959), p. 42. Cf. also Tod, Travels, p. 264.

[39] Cf. Forbes, Ras Mala, pp. 353-55; Gordhandas Nagardas Mehta, Saurashtra Itihaas Darshan (Palitana: B. P. Press, 1936); and Watson, ed., Statistical Account of Bhavnagar on the one hand with Shri Yaduvansh Prakaash and Watson, ed., Statistical Account of Nawanagar on the other.

[40] For a fuller account, see M. S. Commissariat, A History of Gujarat (Bombay: Orient Longmans, 1957), II, 56.

to gain nomination of favorite sons as appointed heirs were furious. Intrigue in the harem continued as a common feature (perhaps even increased) after the British arrival and is documented extensively in British administrative accounts.[41]

Another group in the city which drew its power from association with the ruler was the administrators and advisors to the ruler--a very mixed group and an indispensable one. In addition to civil administration they also had charge of the state's armies until the British assumed all military authority. The diwan of a major state commanded thousands of troops and was assigned revenues of various districts with which to pay them.[42]

When the British became more active in revenue administration in the 1860's, they attempted to cut back on the number of tax collectors and administrators employed by the princely governments, but without success. The revenue system was so complicated that the new rulers could not even get through the work of collection without assistance, much less carry on judicial functions at the same time. Each village often had its own tax policies, and these differed even from crop to crop.[43]

Among the administrators, one caste group, the Nagars, stood out. "The Nagar community," wrote a British official in 1842, "is very powerful in the peninsula; they are by profession a <u>corps diplomatique</u>, and devoted to the arts of government. Their principal residence is Junagadh, but there are many families at Nawanagar, Bhavnagar, and other large towns."[44]

A biography, written by a Nagar, of Gaurishankar Udayashankar Oza, the great Nagar diwan of Bhavnagar in the early and mid-nineteenth century, gives a capsule history of the rise and importance of the Nagars. The author writes that as early as the Vallabhi dynasty, ca. 470-788, Nagars were put in charge of some of the Cambay possessions. They became increasingly concentrated in statecraft and migrated to Saurashtra and Kutch as well as throughout Gujarat. They were given many villages in return for service. Their skill was based on their knowledge of language, history, and even poetry. They were go-betweens

---

[41] See especially Jacob, <u>Western India</u>, pp. 22-55 and Gense and Banaji, VIII, 329-49.

[42] <u>Tarikh-i-Sorath</u>, p. 176.

[43] National Archives of India (NAI), Western India States Agency File (WISA), Vol. VII on Gondal Agency.

[44] <u>Selections</u>, XXXVII (New Series), 29.

and political managers.[45]

The Nagars throughout the peninsula, at least through the early years of British rule, acted together to maintain their power at the expense of the ruler. The Resident at Baroda, Major Wallace, wrote in 1861, "Some of the chiefs of Kathiawar [are] having their ministers forced upon them by the Nagar bureaucracy of the province. A string of names of such chiefs was given me with the expressive addition that each was in this regard a 'bundeevan' or a prisoner in his own palace."[46] Other caste groups, notably bania groups, usually thought of as merchants, also participated in advisory roles in Saurashtra. Hasmukh Sankalia, the historian and archaeologist of ancient Gujarat and Saurashtra, reports that from medieval times, Modh and Porvad Banias assisted in royal government.[47] The example of Mahatma Gandhi's Modh Bania family serving as diwans in various Saurashtra states is thus just one case in a not uncommon pattern.[48]

The Khawases, or illegitimate children of the rulers, were another group which occasionally rose to places of prominence among the raja's advisors. One of the most famous was Meraman Khawas. Meraman began as a servant to one of the wives of the Jam Lakhaji of Jamnagar. He usurped power and remained in de facto charge until the death of Jam Lakhaji in 1768 and throughout the reign of his successor, Jam Jasaji.[49] "The Jam Jasaji was Jam in name

---

[45] Kowshikaraam Vighraharaam Mehta, ed., Gaurishankar Udayashankar Oza (Bombay: Times of India Press, 1903). For a comparison with a similar group in South India, see Robert Frykenberg, Guntur District (Oxford: Clarendon Press, 1965).

[46] Affairs of Kattywar, Part II, p. 47. I have no further publication data on this volume. I found it in the District Library, Rajkot, along with other books deposited from the former archival deposits of the Rajkot Residency. The volume reprints a series of letters and reports on the status of Saurashtra about 1859-1861 as changes in administrative status were considered by the Bombay and central governments.

[47] Hasmukh D. Sankalia, The Archaeology of Gujarat (Bombay: Natvarlal and Co., 1941), p. 210. The presence of the Saurashtra banias in both trade and statecraft challenges D. R. Gadgil's assumption that "In Hindu Society there was no mobility between the merchant-trader classes and the military, priestly, and ruling administrative classes," Origins of the Modern Indian Business Class, an Interim Report (New York City: Institute for Pacific Relations, 1959), p. 23.

[48] Mohandas Karamchand Gandhi, Autobiography (Boston: Beacon Press, 1957), p. 3.

[49] Kincaid, Outlaws, p. 89.

only, as he was kept by Meraman and Bhowan Khawas, the karbharis, under surveillance, like a parrot in a cage, whilst they reigned in [Jam]nagar according to their pleasure and collected much gold and silver."[50] Meraman died a natural death in 1800 having survived various attempts on his life and having repulsed many concerted attempts by neighboring states to release the Jam Saheb and destroy Meraman.[51]

Similarly, in Porbandar, khawases rose to great importance, threatening the power of the ruler. In 1810, both the ruler of Porbandar, the Rana Haloji, and his son, Prithiraj, were advised by khawases. As father and son opposed each other bitterly, the son withdrew from the capital to the fort of Chhaya, a few miles away. The father claimed that he was so frightened of his son and his son's khawas advisor that he had to spend an extra Rs. 80,000 per year, about 40 per cent of his usual annual revenues, to hire extra armies. This costly enmity was generally attributed to the influence of the khawases, especially the advisor of the son.[52]

A fourth very important group in the urban power structure was the banias or merchants. Their base of power was control of capital and this gave them some independence of the court. Michael Pearson, in his dissertation on the merchants of sixteenth century Gujarat, documents this in depth for the mainland.[53] This independence was especially marked among traders who dealt in long-distance trade and trade by sea rather than those who traded mostly within the confines of a single raja's domains.[54] Data both on sea trade[55] and on the prevalence of piracy[56] suggest considerable coastal and long-dis-

---

[50] Tarikh-i-Sorath, p. 168.

[51] Shri Yaduvansh Prakaash, pp. 267-86.

[52] Gense and Banaji, VIII, 359-62.

[53] Pearson, "Commerce and Compulsion." Also see his "Political Participation in Mughal India," Indian Economic and Social History Review, IX, No. 2 (1972), 113-31.

[54] For a similar kind of independence in the economic sphere in nineteenth century Poona, cf., D. R. Gadgil, Poona: A Socioeconomic Survey (2 vols.; Poona: Gokhale Institute of Politics and Economics, 1945 and 1954), II, 38-41.

[55] Bhavnagar, the main port of the peninsula, reported imports valued at Rs. 1,294,427 and exports of Rs. 2,296,456 in 1799-1800. Watson, ed., Statistical Account of Bhavnagar, pp. 14-15.

[56] Tod, Travels, pp. 429-43.

tance trade. In 1789, Bhavnagar port was exporting one-third of all of Gujarat's cotton, 3,000 tons of a total 9,000.[57]

On land as well as sea, the chief source of the merchants' power was control of capital. A very unflattering picture of their exploitation of the countryside from their urban base is given by Forbes in his Ras Mala: "A wanee commencing life spends his time partly in a large town and partly in some remote country town."[58] He borrows a few staple household needs in the former to sell in the latter. He is repaid with grain in the latter which he sells to the former. He buys cheap and sells dear. Soon he has recouped his expenses and builds up his capital.

As his lending business expands he begins to lend for field animals, marriages, and major expenses, charging great interest. If the farmer wishes to go to town himself to do his own buying, the bania goes with him, assuring him that otherwise he will be cheated. Thus the bania further cheats the farmer in price and in charging for his services.

Finally, with mounting interest charges, the moneylender has trapped the farmer. The moneylender ceases to flatter the farmer and begins to lord it over him. The bania closes his village shop and moves permanently to the city. When he comes to the village for business he forces the farmer, who is now deeply in debt to him, to put him up and also to pay his special expenses. The moneylenders, too, apparently formed a hierarchy with the most successful financially living in the capital towns and carrying on business with the rulers; the less successful in small towns and villages.[59]

How active were the banias politically? Ranchodji Amarji, son of the Junagadh diwan and surely in a position to know, said that generally the banias in trade did not seem active politically. Many of the wealthy put their money into religious shrines, and into roads to pilgrimage places which in Saurashtra were often out-of-the-way.[60] Forbes, on the other hand, points out that the moneylenders ensnared large landholders, girasias, and rajas into debt in just the same fashion as they ensnared the simple farmers; only the scale was larger.[61] Major Walker, describing the pargana of Dholka just outside the eastern

---

[57] Nightingale, Trade and Empire, p. 29n.

[58] Forbes, Ras Mala, p. 547.

[59] Kathiawad Gazetteer, p. 170.

[60] Tarikh-i-Sorath, pp. 28-29.

[61] Forbes, Ras Mala, p. 547.

border of Kathiawad, supports Forbes' view. He writes that the town dwellers, "kusbatees," assisted the local ruler, the Gaikwad, in collecting revenues, and in return the Gaikwad respected their privileges.[62] More, "The greatest part of the Sirkar Zemin [literally, the government land] has been sold and mortgaged to the kusbatees and girasias."[63]

An example from the Jodiya Balamba taluka in Nawanagar during the early years of British control indicates the different kinds of control which could be held over land. The ruler and later his rebellious khawas servant both claimed official control over the land, its people, and its revenues. They backed this claim with military force and expected it to be recognized. When the ruler failed to pay the revenues demanded by the British, the revenues of the land were mortgaged to a bania. Although he thus gained control over land revenues, he did not gain any legal hold over its inhabitants, nor any recognition as a ruler. He controlled no military force. When the debts were paid out of the annual revenues, control over the land income returned to the ruler.[64]

It appears that the banias wanted primarily to amass capital, to conduct their business with a minimum of obstruction, and to enjoy the fruits of their wealth. The rulers, on the other hand, were less concerned with capital and more with control of land, military forces, and the sense of power which came with giving orders.[65] In large measure these two interests could co-exist with the prince supplying the muscle and the bania the financial talent of the rajadom.

---

[62] Selections, XXXIX, 14.

[63] Ibid.

[64] "Kovass Sagram, a servant of the Navanagar Darbar, had obtained possession of these Mahals, and settled for them as an independent Chief; but, in consequence of his being engaged in a rebellion of the Mashatty Arabs in the Jam's service, a force was sent against him in 1815, and the territory was restored to the Jam. Being required to furnish security for the repayment of the military expenses, he made it over in mortgage to Shet Sunderji Sewji, who held it on these terms until A.D. 1825 when it was included in the farm of the Nawanagar Taluka. On the dissolution of that engagement it was placed in attachment, on a plea of the former mortgage, pending an adjustment of accounts; but no ostensible balance appearing against the Jam, and the Shet refusing to submit his claims to Panchayat, it was delivered over to the former at the commencement of 1830, on certain conditions respecting the claims of the Sunderji family specified in a writing then given by him to Government." Selections, XXXIX (New Series, 1894), 119-21.

[65] For a view of the personality of a contemporary Rajput in a ruling situation, cf. Gittel Steed, "Notes on an Approach to a Study of Personality Formation in a Hindu Village in Gujarat," in Village India, ed. McKim Marriott (Chicago: University of Chicago Press, 1955), pp. 102-44.

Banias did not give directions for military rule and rajas did not interfere with the banias' capitalistic efforts. Rulers of men and land and rulers of capital eyed one another warily but did not challenge each other's culturally sanctioned roles. They did, however, occasionally encroach on one another's functions.

For merchants to be influential in other spheres they had to organize. And a fifth group in the urban arena was, in a sense, an extension of the merchants in a formal or informal corporation known as a mahajan, often translated as a guild. I see the mahajan as a distinct group, for often it acted not only for economic goals,[66] protecting the collective interests of merchant groups, but also as a religious pressure group. The Hindu banias of Saurashtra are, by sect, usually Vaishnava Hindus or Jains; both groups share, in addition to other ideologies, a strong belief in non-violence and vegetarianism.[67] British officials occasionally commented on the difficulty of obtaining meat because of the vegetarian religious practices of the mahajans. Perhaps most striking was the effect of the Bhavnagar mahajan's influence on the Maharaja of Bhavnagar around 1815. They told the Maharaja that cow-slaughter in his state was a disgrace to him, and he followed their advice by executing two cow-slaughterers. The British found this repugnant, lowered the Maharaja's status, and took away his control of 116 villages, about one-fifth of his domain. Pressure from the mahajan, as this example illustrates, could have important repercussions in state affairs.[68] The link between occupation and religion could be divisive, too. Clearly it separated the interests of the very sizable Muslim trading groups from those of the Hindus and further subdivisions could be found within these various religious categories. The mahajan served as a religious-

---

[66] I do not have specific examples of mahajan pressures on Saurashtrian rulers in pre-British days. References in the history of Bhavnagar State during the Independence struggle do indicate that such pressures had been common. See Bhavnagar Praja Parishad Trijun Adhiveshan (Botad, 1928), a pamphlet distributed at the Bhavnagar People's Conference, Third Session, Botad. Copy in Barton Library, Bhavnagar. For mahajan pressures in the neighboring region of northern and central Gujarat, see B. G. Gokhale, "Ahmedabad in the XVIIth Century," Journal of the Economic and Social History of the Orient, XII (April 1969), 187-97.

[67] For reference to the religious interests of business people and their consequent entrance into politics in north India, see C. A. Bayly, "Patrons and Politics in Northern India," in Locality, Province, and Nation, ed. John Gallagher, Gordon Johnson, and Anil Seal (Cambridge: University Press, 1973), pp. 29-68.

[68] Watson, ed., Statistical Account of Bhavnagar, p. 39. See also Mehta, ed., Gowrishankar Udayashankar Oza, for a fuller account.

commercial corporation; the two components were not separable.

A sixth group in the cities which could become politically important was the standing mercenary army.[69] Often the looting so characteristic of the Saurashtra countryside seemed simply a method of paying the armies by allowing them to loot in their own states or by encouraging them to plunder a neighbor.[70] In battle, the soldiers had more than once been known to switch sides as, for example, Loma Khuman did in taking twelve thousand horses from Jam Sataji over to the Mughal Viceroy as the Mughals captured Saurashtra in the crucial battle of Bhuchar Mori in 1591.[71] In peace time, as well, the troops could be a threat. In Junagadh in 1761 the Arab armies of the state imprisoned the Nawab Saheb Mahabat Khan, even appointing a successor, until agreements were reached paying off the Arab armies and freeing Mahabat Khan.[72] When the British began actively to administer Saurashtra affairs, one of their first concerns was to disarm, and where possible, deport the mercenary soldiers. With no fighting to do, these mercenaries posed an especially serious threat to their own states and to other states as well.[73] A weak ruler, lacking the resources to pay his troops, found himself at their mercy.

Sometimes the Pax Britannica led to striking new balances between the mercenaries and the other groups; this suggests that the transition between peace and war could sharply alter the political structure in earlier times as well. Tod recounts that the chief of Palitana had hired Arab mercenaries for his defense. When the British imposed peace, these troops became his chief danger. He therefore engaged a banker to whom he mortgaged all his estates, including the pilgrim tax. He reserved for himself only Rs. 40,000 per year and the banker advanced the sum necessary to pay off the turbulent Arabs.[74]

Another group of mercenaries who posed a threat during peace time was the Mianas of Malia in northern Saurashtra. They had come to Saurashtra from Sind under the chief of Malia in the eighteenth century. They fought with him

---

[69]Selections, XXXIX, 292.

[70]Shri Yaduvansh Prakaash, passim.

[71]Desai, Saurashtrano Itihaas, p. 532.

[72]Tarikh-i-Sorath, pp. 144-45.

[73]Dhanjishah Hormasji Kadaka, ed., The Kathiawar Directory (rev. ed.; Rajkot: Damodar Govardhandas Thakkar, 1886), pp. 411-13 and 493-94.

[74]Tod, Travels, p. 300.

against Morvi State, and when the fighting was over they stayed on in the region living by plunder. Under British rule they were classified as criminals and made to report to police headquarters twice daily for roll call.[75] These later experiences suggest that the mercenary soldiers, particularly the immigrants, posed a threat to their own employers in times of peace and presumably acted as a constant goad to waging war.

A seventh group, but one that seems not to have been important in the regional power equation, was the artisans.[76] Although they might combine to attain their own goals, and although at least in some parts of India they might be able to enforce discipline within their ranks,[77] they had little concern with politics outside of issues of immediate interest to themselves in trade. Trade, however, was not under their control but was dominated by banias.

Parenthetically, and perhaps surprisingly, nowhere does the literature suggest that the Brahmins as priests exerted special influence in public affairs. Brahmins and Nagars--often considered a subcaste of Brahmins--did enter administrative services, but this was an alternative to the priesthood rather than a fulfillment of it.

In sum, if affairs of land tenure and the mansabdari system of administration have kept scholarly eyes focussed on the countryside and the landholders on the one hand and on the court and its nobility on the other, a more balanced picture must include urban groups who, though not controlling land, had skills, capital, organization, and/or personal relationships with the rulers which made them powerful. They, too, contested for various kinds of power, sometimes outside the professional concerns of the ruler, sometimes tangentially to them, and sometimes in direct consonance or conflict with them. To understand the structure of rule in the province, one must understand the interactions of these groups.

The Interactions of Urban Groups

In effect, two types of power and rule existed. One, based on control of territory (and followers necessary to hold the territory), was dominated by

---

[75] Kincaid, Outlaws.

[76] Cf. Tarikh-i-Sorath, p. 244.

[77] Cf. E. Washburn Hopkins, India Old and New (New York: Charles Scribner and Sons, 1902) and Cohn, "Political Systems in Eighteenth Century India."

princes. To the extent that other groups supported the landholders, they too were drawn into this network, binding themselves to territorial power bases. The other type was based on control of capital, skills, knowledge, and sometimes on access to the ear of the ruler. The groups of people who sought wealth and political influence often required territorial mobility. Their freedom to move gave them leverage.

Contestants whose power rested on control of land became tied to land; it had to be defended against others for whom territory was also power; it would have to be expanded if more power was sought. Since land was a fixed commodity, and since even the more elastic rights in land were ultimately limited by the productivity of the soil, the gains of one contestant had to be at least in some measure the losses of another. Three possibilities could result: a single power could become so pre-eminent that no one dared challenge it seriously; agreements on an accepted balance of power could be negotiated; or war could ensue. Even under the Mughals, the central power had not been able to enforce a balance of power effectively in Saurashtra. With the Mughal decline and the failure of the Marathas to replace them as a force within the peninsula, even the possibility of a balance enforced from outside vanished. Agreements among indigenous powers were attempted in some regions, wars in others. In the early years of the nineteenth century, this search for a decisive military-political stability in Saurashtra--as in other parts of India--brought British armed power and administration into the region. Bhavnagar had early entered into temporary alliances with the British to put down piracy.[78] Other smaller states meanwhile offered to cede to the British shares of their revenues for protection against the Maharaja of Bhavnagar.[79]

While landholders played out their roles in a territorial system anchored to their land, other participants were not so geographically bound. Administrators, merchants, soldiers, artisans, farm laborers, and even farmers could and did move from place to place, from one rajadom to another, in search of improved opportunities. We have already discussed above the activities of some of these groups in seeking to influence the rulers; here we detail their alternative course, emigration.

Ranchodji claims that when his father lost the diwani in Junagadh as the result of court intrigue, he was offered positions by twenty-eight governments,

---

[78] Forbes, Ras Mala, p. 418.

[79] Nightingale, Trade and Empire, pp. 211-12.

both within Saurashtra and outside, including Nawanagar, Surat, and the Portuguese at Diu.[80] Ranchodji also noted that occasionally a diwan would lead a revolt against his ruler, as Kalyan Sheth did against the Nawab of Junagadh in 1789 by fleeing to Kutiana.[81] So common was the movement of diwans that Mahatma Gandhi, in describing the ministerial work of his grandfather in the courts of Saurashtra, notes that even when he moved from a position in Porbandar to one at Junagadh, he re-affirmed a loyalty to Porbandar; most diwans evidently felt no such sense of loyalty to territory or to rulers.[82]

Mercenary soldiers, by the nature of their work, were of course free agents and highly mobile.

Merchants, too, migrated to gain advantage. This seems especially true of long-distance traders.[83] Moreland, discussing the Muslim traders of India's west coast ports, notes that they

> acquired a privileged position owing to the fact that they could make or mar the trade of a particular port; merely by staying away they could ruin the local merchants and, what was probably more important, they could cause serious loss to the administration which depended on the port dues for a large part of the revenue, or to the Governor, if he had farmed the customs for his private benefit.[84]

While Moreland's study deals with the period around 1600, it accords well with the data on emigration of merchant castes even in the early twentieth century; by 1911, 10 per cent of Saurashtra-born male merchant-caste members had emigrated to Bombay City.[85] The continuous competition for traders between Bhavnagar, Gogha, and the other seaports of the Gulf of Cambay during the eighteenth and early nineteenth centuries illustrates the persistence of the pattern.[86] Indeed one of the trade techniques of the British East India Company

---

[80] Tarikh-i-Sorath, pp. 157-58.

[81] Ibid., p. 89.

[82] Gandhi, Autobiography, p. 3.

[83] A surviving autobiography of a seventeenth century north Indian bania dramatically illustrates the importance of mobility to the survival, much less the prosperity, of an Indian merchant. Ramesh Chandra Sharma, "The Ardha-Kathanak: A Neglected Source of Mughal History," Indica, VII, No. 1 (March 1970) and No. 2 (September 1970), 49-73 and 105-20.

[84] W. H. Moreland, India at the Death of Akbar (Delhi: Atma Ram and Sons, 1962), pp. 186-87.

[85] Census of India, 1911.

[86] Forbes, Ras Mala, p. 420; G. N. Mehta, Saurashtra Itihaas Darshan, p. 54.

in Saurashtra before they assumed political control was to establish protected ports which would attract merchants from the princely areas.[87]

Even farmers migrated. Folk songs reported in the Ras Mala indicate the constant battle between the kunbi peasant and the landlord over taxes. The landlord usually tried to prevent the kunbi from fleeing, but despite his efforts, and despite the confiscation of the real property he left behind, the kunbi did flee to states where land was more abundant and landholders less rapacious.[88]

Flight of the peasantry had a long history in Saurashtra. In 1642, under the Mughals, even the great Azam Khan was removed as Governor of Gujarat because his pacification of Saurashtra had been so oppressive as to drive peasants from the region.[89] Jacob reported in 1842 that the threat of desertion by the peasants acted as a check on despotic princes. Especially if the patel, a leading influential cultivator, were to leave the village, a proportion of the skilled and influential villagers were likely to move with him, as their affairs were closely intertwined.[90] Many parts of Saurashtra had open land to which new farmers were welcome.[91]

Bahaarvatiyos, going into armed opposition to their rulers, also often left their villages for another prince's domain where they would establish their base of guerrilla operations.

Physical mobility was often associated with occupational mobility as well. Some examples have been cited above of people changing occupations in mid-career. There are further examples of links between this occupational mobility and geographical mobility:

> In Samvat 1874 [1816 Christian Era] Sheikh Amrullah who was originally an indigo dyer, and who had been allowed by the deceased Diwan Saheb Amarji to establish himself in the town [Junagadh] and who by his trade in rich Ahmedabadi cloths and all kinds of stuffs, gradually wormed himself into the court of the Nawab Saheb, and into the favor of the Masaheba Raj Kumar, succeeded at last in attaining the rank of companion (Musaheb) to the Nawab Saheb.[92]

---

[87]Nightingale, Trade and Empire, pp. 175-235.

[88]Forbes, Ras Mala, pp. 543-45.

[89]Mirat-i-Ahmadi: A Persian History of Gujarat, trans. M. F. Lokhandwala (Baroda: Oriental Institute, University of Baroda, 1965), p. 189.

[90]Selections, XXXVII, 12-14.

[91]Ibid., XXXIX, 203.

[92]Tarikh-i-Sorath, p. 216.

Similarly the history of Bhavnagar recounts that Dhanji and Manji Mehta left their ancestral land in Mahuva, settled in Sihor, then the capital of the Gohel Rajputs, to supply provisions for the troops. Seeing their skill, Maharaja Bhavsinhji employed them as his private administrators. They were appointed to carry on negotiations for peace with the Marathas. The story of the rise of these Kapol Banias found its way into the books of the Barots, who usually detail only the exploits of Rajput military history, because they were treacherously killed while on a mission of state to the Marathas.[93]

Competition among the small states of the fragmented peninsula provided marked advantage to men with physical mobility. As Alfred Lyall remarked of the similar, Rajput-dominated states immediately to the northeast of Saurashtra:

> In Rajputana alone there do actually exist the natural institutions which, in various forms and stages, have checked and graduated the power of sovereigns all over the world. The incessant bickering and contests between encroaching Chief and jealous kinsmen; the weak central power, the divided jurisdictions, the obstinacy with which a man of high birth insists on the proper punctilio to be reciprocated between himself and his Chief--all these are tokens of a free society in the rough.[94]

Town and Village

Physical mobility brought a measure of power to rural groups as well. The opportunity and encouragement for rural mobility derived from the carrot of available virgin land and the stick of total control over real assets maintained by the rulers. As to the availability of land, an average of at least 75 acres seems to have been available per farm family. I extrapolate this figure from the availability of approximately 25 acres per farm family in a population of 6,000,000 at the time of Independence to the population of below 2,000,000 at the time of the British assumption of power. Seventy-five acres is beyond the capacity of individual families to farm, and indeed, records at the time of British settlement indicate that rulers did attempt to attract new settlers to their empty lands.

An invaluable report by the British officer George LeGrand Jacob in 1842 supplies a fascinating record of the process of village formation. First came the preparation of the area and the invitation to potential settlers from the ruling Chief:

---

[93] G. H. Mehta, Saurashtra Itihaas Darshan, p. 55n.

[94] Sir Alfred Lyall, Asiatic Studies (London: John Murray, 1907), I, 262.

> The first process is an examination of the ground by the Chief in person or his deputed agents, and on the site being fixed, he gives out publicly his intentions; hereupon men who fancy they can better their condition by change, and who can command from two to a hundred ploughs, proceed to make their terms, which vary according to the character of the Chief, the quality of the soil, etc., but principally only as to the amount of profit for the first two or three years, after which though payment is made under different heads more or less varying, the general result differs but little, varying from a third to half of the produce in kind, with a proportionate increase in fixed money tax. Generally the cultivators receive for the first year of occupancy the whole of the produce, the second year a small proportion is assigned to the Chief, and the third year, unless the ground had required great outlay for clearing, he receives his full rights, as fixed by the deed of agreement passed to the Patel or Patels who have brought the rayats to him.[95]

But the rulers did not give permanent rights in land or fixed property to settlers. "The cultivators," Jacob noted, "have no property in the soil, which is exclusively that of the Chief."[96] Even in the most enlightened of Kathiawar's major states, Bhavnagar, the raja retained all rights to the land in all villages and even in many towns. Improvements on the land, including new houses built, belonged to him and could not be sold. The result, even well into the British period, was to leave the countryside--as opposed to cities where inhabitants did have rights of ownership and sale--underdeveloped. Even as late as 1872, "omitting Bhavnagar, Sihor, Kundla, Rajula, and Mahuva [the major towns], the number of houses of brick or stone is only 660 to 98,770 built of mud."[97]

Among the settlers was a diverse occupational spectrum. Jacob estimates that for every one hundred farming families, there were two families of carpenters, sutars; one or two blacksmiths, lohars; two tailors, darjis; two potters, kumbhars; one or two leatherworkers, mochis; two barbers, hajams; four sheep and goatherds, bharvars; eight or ten dhers, men who act as curriers and perform the rough work of the village; three or four banias; and eight or ten pasaitas, armed watchmen. Over time, the affairs of these people become intertwined--a process presumably intensified by the system of reciprocal services and payment in kind rather than in cash--and if the leaders chose to migrate, most of the rest of the community might follow:

> All these classes hang together and their dealings become so mixed up with one another, that when a Patel or leading cultivator of influence, quits one place for another, a proportion of these will always accompany him: there will generally be two or three leading men among them, who act as leaders

---

[95] Selections, XXXVII, 12-13.

[96] Ibid.

[97] NAI, WISA, 1872, Vol. 2.

of the rest--the bania, who advances grain for seed, and money for bullocks; the man who, by money or influence, can command the greatest number of ploughs; and the most skillful of the artisans.[98]

It is these leaders who make the agreement with the Chief:

> All these customs, accidents, and risks, the Patel or Patels calculate on ere they take up their residence in a new quarter; but once having agreed to the terms offered, they receive a turban in token of engagement from which they cannot draw back without exposing themselves to a fine entered in the agreement. They then proceed to form the village in the allotted quarter.[99]

Later, others may come to join the community.

The political tie between village and capital was thus the direct, personal tie between the Chief and the village leaders. If the Chief was the ruler of only a few villages, and had not adequate resources or desire to build a major capital, he might live in the village himself. Alternatively, in order to establish his control and presence, and simultaneously to provide payment in exchange for local supervision, he would designate a representative, usually a blood relative, to live in the village. The power of the Chief or his representative was symbolized by his residence in the village. The house of the proprietor would be located in the center. It might be three stories or more in height, the tallest structure in the village. If there were more than one shareholder, each had a house. Houses of proprietors were surrounded by a courtyard in which rows of huts accommodated the domestics and animals. Around the main house were less imposing houses of relatives, and the chief streets of the town radiated out from them to the town gates.[100]

In exchange for the one-third to one-half of the produce in kind and some fixed cash fees which the cultivators paid to the Chief, what did they receive in return? A right to use the land, an expectation that the Chief would keep order and defend the villagers from outside attack, and an expectation that the Chief would attempt to keep a food supply available against famine years by storing surpluses in good years and importing. This was the expected return for revenues. Since this was approximately the standard bargain across the peninsula, and indeed across the subcontinent,[101] it apparently seemed acceptable to contemporary sensibilities.

---

[98] Selections, XXXVII, 13.

[99] Ibid., p. 12.

[100] Kathiawar: Gazetteer, p. 170.

[101] Cf. Habib, Agrarian System; Moreland, Agrarian System; and Sarkar, Mughal Administration.

The economic link between the village and the city seems to have been the bania. The only two evaluations of his role which I have seen are contradictory. A. K. Forbes' discussion cited above sees the bania as charging excessive interest and stationing himself between the farmer and the market, by deception if necessary, so as to take an unnecessary service charge and an exorbitant profit from the transaction. According to Forbes, the bania's main goal seems to have been to exploit the villagers until he earned adequate profits to leave for the city. The moneylenders, like the rulers, apparently formed a hierarchy, with the most successful living and working in the cities and maintaining a chain of links with those in the smaller towns and villages. It is unclear whether they were linked by blood, but a caste tie is certain.

Jacob's view of the bania is markedly different. He presents the moneylender as a man of the village and responsive to its needs, a leader who aids the farmers by advancing vitally necessary grain for seed and money for bullocks. In the absence of further data, both these views are tenable. Only data, unlikely to be uncovered, on the actual profits taken by banias, interest rates charged, rates of mobility from village to town, and attitudes of various other classes toward them, would help clarify the situation. In addition, to the extent that the economy was largely non-monetized and in which even revenues were collected in kind, the economic links between village and town may be minimal.

Alternative City-State Policies

Plainly a state might benefit by attracting skills and capital of foot-loose potentially productive people. Bhavnagar seems to have been the most active state in exploiting this possibility in the eighteenth century. Under two great rajas, Bhavsinhji who in 1723 founded the city which bore his name, and later his grandson, Wakhutsinhji, Bhavnagar set on a path of commercial expansion. At the time of Bhavnagar's founding, Gogha port, about ten miles away, dominated the export trade of northern Gujarat, Malwa, and Rajputana. As the Mughal power began to fail, however, and the Marathas grew powerful, these two outside powers put Gogha under their joint rule, appointed rapacious officials who imposed excessive taxes, and depressed trade. Unhappy in Gogha, the traders came to Bhavsinhji and requested facilities for establishing their export business at Bhavnagar. Bhavsinhji seized the opportunity by providing facilities, low export duties, and special considerations.[102] He protected wrecked

---

[102]G. N. Mehta, Saurashtra Itihaas Darshan, p. 54.

ships and returned stranded vessels to their owners; at Gogha, such ships were farmed as a source of revenue.[103] The Gogha merchants began to migrate to Bhavnagar.

Wakhutsinhji, the grandson, not only expanded the trade of Bhavnagar, but he also captured Talaja and Mahuva, and developed the latter into a flourishing port, again by attracting new merchants.

> Under these auspices, Bhavnagar became the channel of the import and export trade of Gujarat, Sorath, and Marwar, and the encouragement which merchants received induced many opulent people to settle there, while the neighboring port of Gogho, with the advantages of a much more convenient harbor, soon fell into decay.[104]

Tod described the success of Bhavnagar's policy, perhaps tongue in cheek: "As we entered the city there was nothing to call for particular notice, excepting the crowds of wealthy merchants traversing the streets, from whom as the poet Chund says, 'cities derive their beauty'; and in this point of view, Bhaonuggur was certainly beautiful."[105]

The Bhavnagar rulers understood that not only trade but also administration of high caliber was necessary and they recruited assiduously. Data on migration of Nagars indicate success in this policy as well.[106] A further, more specific example was Bhavsinhji's enticing of Sorab Khan, influential even in Delhi, into his government after he was expelled from the government of Surat in 1732.[107] Such far-sighted policies made Bhavnagar in power, wealth, and political finesse the strongest state in Kathiawar.[108]

Not all rulers were so wise or so powerful. The Tarikh-i-Sorath indicates that emigration of town population often accompanied a change in rule. Nagars and Lohanas, a trading caste, might flee if they feared the policies of a new ruler.[109] The urban arena was both their point of emigration as well as

---

[103] Selections, XXXIX, 163.

[104] Forbes, Ras Mala, p. 420.

[105] Tod, Travels, p. 260.

[106] Watson, Statistical Account of Bhavnagar, p. 11.

[107] Ibid., p. 24.

[108] Wilberforce-Bell, History of Kathiawad, p. 127. For an example from south India of a new state administration building up its capital city by attracting migrants, cf. Karen Leonard, "The Hyderabad Political System and Its Participants," Journal of Asian Studies, XXX, No. 3 (May 1971), 569-82.

[109] Tarikh-i-Sorath, p. 92 and passim.

their chosen destination. At times of weak rule, all groups might flee. In 1810, Porbandar was caught in a vicious fight between the ruler and his son. Security was uncertain and plunderers invaded from outside. The result was described by an English officer:

> Many of the merchants of Porbandar . . . had in consequence of these disturbances been obliged to retire to Mangrol and Veraval. Many villages have become waste, and some of the principal do not now yield 1% of their produce. A melancholy and striking instance of this was observed in the road by which I travelled to Porbandar. From Kutiyana to Porbandar is 27 miles, and the country generally fertile. Kandorna and Ranavan were the only two inhabited villages passed in the road. They formerly employed 600 ploughs, but at the present time there are not more than 80 in use.[110]

Conclusions

The cities of Saurashtra at the beginning of the nineteenth century were primarily ruling centers. Their founding and maintenance depended basically on a political-military ruler choosing them as his capital. In this respect they were surprisingly volatile, for the ruler's whim could bring a new city to existence or empty an old one. Of the five capital cities which we shall be examining in detail, only Junagadh was ancient. Of the others, Nawanagar was established in 1560, Bhavnagar in 1723, and Rajkot and Gondal in between. They were essentially new cities.

Most urban groups in the region were not strongly attached to place. The ruler was; his fortress was his base of power both for controlling land and for facing the other, often hostile and competitive, rulers who divided the peninsula into petty rajadoms. Professionals, businessmen, artisans, even farmers, however, had a marked degree of mobility. The vast number of competitive capitals in a small area gave opportunity to those urbanites willing to migrate, and the relatively sparsely settled land offered similar prospects to farmers.

Within the city, relationships were flexible. Though the political-military rulers were nominally in charge, a number of powerful groups raised their own demands and forced negotiations. Each group had its own sphere of interests--rulers, their relatives, merchants, moneylenders, professionals, artisans, soldiers. To some extent these interests were separable, as reflected in the division of the city into mohullas, where groups lived separately from one another. But to some extent interests converged and here the rulers often had to accept compromises or face the possible loss of valuable subjects. The

---

[110] Gense and Banaji, Gaikwads of Baroda, VIII, 362.

same was true of the city's relationship with the countryside; excessive severity might well result in depopulation. Rulers were constrained in their measures for an empty countryside was of little use.

Clearly different cities had different experiences. Certainly no city housed an egalitarian society nor promoted equality with the countryside. Extremes of wealth and poverty were the norm; even Bhavnagar which clearly promoted economic development through trade held the countryside under its thumb and did not allow, for example, private ownership of rural land. On the other hand, few cities could afford to be excessively harsh; the desertion of Porbandar by its subjects was an example of the difficulties a rapacious ruler met.

Most remarkable, however, was the enormous volatility of the city-states. They rose and fell, they changed policies, they forged shifting alliances both internally among their own elites and externally with other city-states with surprising speed. No single pattern characterized all the cities of the peninsula at any one time nor any one major city through the entire period. Shifting balances of power and opportunity were the normal pattern.

CHAPTER II

THE EFFECTS OF BRITISH POLICY ON THE URBAN SYSTEM OF SAURASHTRA

> In short, the Rajpoot vaunts his aristocratic distinction derived from the land; and opposes the title of "Bhomia Raj" or government of the soil, to the "Bania Raj" or commercial government, which he affixes as an epithet of contempt to Jeipoor: where "wealth accumulates and men decay."
> --James Tod, <u>Annals and Antiquities of Rajasthan</u>

Colonial indirect rule in Saurashtra drove a wedge between the native ruler and the native ruled. It created a new class which was tied in style, interests, and geographical orientation outward toward the foreign rulers; conversely it severely weakened the ties between the princes and their own countrymen and "subjects." By elevating the urban-based princes and extending new opportunities to professionals and businessmen, the British widened the gaps between urban and rural societies, levels of political consciousness, and degrees of economic development.

Seeking to govern cheaply and to secure the minimal goals of safety for trade, tribute collection, and law and order, the British chose an enclave strategy of rule which further emphasized the power of cities over the countryside. British Political Agents developed Rajkot Civil Station as their capital and concentrated the centralized bureaucracy there. They built transport networks primarily to link the princely capitals with Rajkot rather than to integrate the countryside generally into a systematic market economy. Areas touched by the trunk roads and the railroads benefitted from the new networks; those not reached did not. Increasing polarization resulted, dramatically increasing urban wealth but not spreading the profits widely.

On the countryside, the landholding girasia members of the ruling lineage, in some respects the key link between the urban citadel and the countryside, lost ground. They could not keep up with the new standards set by the princes and the Pax Britannica precluded their revolting in bahaarvatun. Culturally the girasias remained linked to the peasantry. Neither group seemed to benefit from the urban changes, and the gap between city and countryside grew.

Casting this into the terms developed in our earlier chapter on pre-British rule, the landholders who had previously been reliant on the countryside were now reliant on the British. Political authority and the right to land revenue no longer came from conquest and subsequent bargaining with the agriculturists, but from sanction by the British. The group of reference for the urban landholding class shifted.

Those urbanites who had previously drawn their strength from the ability to move about--the businessmen and the political advisors--still did so; on the one hand they found that the expansion of transport and communication networks gave them more scope for movement, but on the other hand, almost all the key points in the new system were dominated by Britishers. The diversity of markets for the services of these professionals and businessmen was curtailed within the peninsula. Many of them left. Later, however, the transportation and communication networks which had enabled the British to organize and rule served these mobile types in organizing to oppose that rule.

British rule in Saurashtra can be analyzed in three periods: 1) from the beginnings of trade in the late eighteenth century, through the assumption of power around 1820, until the Indian Revolt of 1857-58; 2) from the Revolt until the beginning of a nationwide, organized, influential nationalist movement around 1920; and 3) from 1920 to the granting of Independence in 1947. Significantly, as these dates suggest, the British arrival brings Saurashtra's development into the wider development of the subcontinent generally.

British policy at first, and until 1857, had very circumscribed goals in the peninsula: to protect and facilitate the increasing cotton trade;[1] to suppress piracy;[2] to gain trading rights in Kathiawar harbors;[3] to gather the tribute of

---

[1] Nightingale, Trade and Empire, pp. 175-214.

[2] Gense and Banaji, Gaikwads, VII, 507.

[3] Ibid.

some Rs. 560,000 annually first for the Baroda government and, after 1818, for themselves;[4] to keep the peace;[5] and, to a very limited extent, to combat instances of gross immorality which, in the event, meant the suppression of female infanticide.[6] The British wished to govern minimally and cheaply. As much as possible they wished to maintain the existing system of rule within the peninsula.

The first interest of the British in Saurashtra was the burgeoning cotton trade.[7] In 1784, Dr. Hove, in his mission to seek new sources of raw cotton, described himself as one of the first British representatives in Saurashtra. Only five years later, of the 34,000 candies of cotton exported from Gujarat, Bhavnagar was exporting 12,500.[8] In the opening years of the nineteenth century, Kathiawar was supplying almost half of the cotton exports of northwest India, 40,000 bales of a total of 86,500.[9] The large trade was periodically threatened when the cities in which the British traders worked were threatened in the general combativeness of the area.

The British Resident at the court of Baroda, Col. Alexander Walker, had for several years been eager to move militarily into the region. When the Gaikwad ruler of Baroda invited his assistance in 1807-8 in collecting the annual tribute, Walker accepted.[10] He used this opportunity not only to regularize the collection of tribute but also to end the state of warfare in the peninsula which endangered trade.[11]

Walker moved forcefully to regularize the tribute payments by enforcing a permanent settlement.[12] He had representatives of the local rulers meet with

---

[4] Nightingale, Trade and Empire, pp. 175-214.

[5] Gense and Banaji, Gaikwads, VII, 205.

[6] Ibid.

[7] Nightingale, Trade and Empire.

[8] Ibid., p. 29, n. 10.

[9] Ibid., p. 215.

[10] Selections, XXXVII, 104 and Selections, XXXIX, 267-315.

[11] Ibid., XXXIX, 267-315.

[12] Gense and Banaji, Gaikwads, VII, 565.

him and sign treaties fixing the annual tribute payments for the next ten years and giving de jure recognition to those 292 chieftains who sent representatives.[13] This settlement effectively ended the incursion of the tax gathering mulukgiri armies, fixed the rulers from whom responsible conduct would be expected and enforced, and bound them by treaty and oath from attacking one another. It fixed the jurisdictional boundaries of the internal authorities. In short, it fixed the responsibilities for law and order on specific local rulers and, by recognizing them, validated their claims to specific land.

Walker also signed with the maritime states of Saurashtra agreements on joint action to suppress piracy.[14] He began a campaign (which continued for a half century) to suppress the Jadeja Rajput practice of murdering their infant daughters lest they not be able to marry them off in the hypergamous system which obtained among Rajputs.[15] Walker thus achieved, at least in part, three major British goals: protection of trade; official settlement of political and revenue authority and responsibility within the peninsula; and the assertion of British concepts of morality. These principles persisted as the core of British policy until 1947.

The final accomplishment of the first stage of British rule in Saurashtra came between 1820 and 1822 when the British gained the right over the revenue tribute of Saurashtra for themselves. A combination of wars and agreements concluded peacefully brought to the British the rights to the tribute formerly held by both of the Maratha powers--the Gaikwad of Baroda and the Peshwa of Poona--as well as the right to collect and share in the Zortalbi tribute of Junagadh.[16]

By 1820, the paramount power of the British in Saurashtra was clear. Thus far, however, no new era resulted. British administration was kept intentionally minimal, adequate to protect trade and to secure the annual tribute without inciting any rebellion. The chiefs' jealousy over their powers of internal rule was respected.

---

[13] Selections, Vol. XXXIX gives the entire history of the Walker Settlement as reported by Walker, descriptions of the region and accounts of the agreements reached.

[14] Gense and Banaji, Gaikwads, VII, 597-621.

[15] Selections, XXXIX, Part II.

[16] Gense and Banaji, Gaikwads, IX, 20, 28-78, vii-xv, 274-76 and Selections, XXXVII, 481.

Thereafter, for a half a century, government efforts to promote social and political development ground to a standstill. A. K. Forbes, as acting political agent in 1859-60, summed up the situation:

> The general impression among the people of the province appears to be that Col. Walker's settlement was the fixing point in Kathiawar, not only in political, but also in social matters, and that what is principally to be wished for is that no innovation of any kind . . . should be permitted.[17]

The impetus for changing policies toward the native states followed the great Indian Revolt of 1857-58. The states had remained loyal and, in gratitude, the British chose to make them a first line of support for the Raj.[18] At the same time, realizing that the states had generally received fewer economic and administrative improvements than had British India, the government resolved to expedite their development.[19] In Saurashtra the brief, dramatic period for the initiation of change came during the four years, 1863-67 when Col. R. H. Keatinge served as Political Agent.

Though the British now shifted their policy to emphasize development, they still wished to maintain their earlier principles: support for the political status quo; protection of British financial interests; and inexpensive administration. The result of trying to balance the new and old principles was to limit innovation to inexpensive, non-threatening changes. A few basic strategies were developed.

The first was to increase the efficiency of ruling through the princes by centralizing and bureaucratizing the administration. At the same time, administration was kept structurally indirect, in the hands of the princes, who were prevailed upon to donate the funds needed for most of the new improvements.

Under Col. Keatinge, the British classified the states of the peninsula into seven degrees of authorized jurisdiction from the largest and most responsible in which even capital cases could be tried, down to states so small and impotent that they were allowed to try no civil cases and only minor criminal cases involving punishment up to 15 days imprisonment and Rs. 25 in fines.[20]

---

[17] Affairs of Kathiawad, Part II, p. 2.

[18] Thomas R. Metcalf, The Aftermath of Revolt: India 1857-1870 (Princeton: Princeton University Press, 1964), and Urmila Phadnis, Towards the Integration of Indian States, 1919-1947 (Bombay: Asia Publishing House, 1968).

[19] Bhupen Qanungo, "A Study of British Relations with the Native States of India, 1858-1862," Journal of Asian Studies, XXVI, No. 2 (February, 1967), 251-65.

[20] Wilberforce-Bell, Kathiawad, p. 221. Keatinge also introduced gun sa-

Below these seven levels were a host of still smaller estates covering about 10 per cent of the land area of the peninsula, which retained no civil or criminal jurisdiction at all but were put under direct British control.

The British retained the right in all states, "to interfere at all times when occasion required for the preservation of peace and maintenance of order."[21] They reserved the right to take over the administration of any state temporarily during the minority of a new ruler, a period of state indebtedness, or at the outrage of gross misrule. This right was frequently exercised.[22] In cases of British administrative interference, costs were always borne by the native states.

They founded a few new institutions. The two most imposing and influential of the institutions were the Raj Kumar College opened in 1870 to train the sons of princes for leadership in British specified patterns, and a fund for a trunk road system in the peninsula. Other institutions included male and female teacher training colleges; facilities for vaccina-

---

lutes to Saurashtra in 1865. Cf. also William Lee-Warner, The Native States of India (London: Macmillan and Co., 1910), 375-78 and NAI, WISA, 1878, Vol. II, No. 3A.

[21] William Lee-Warner, "Kathiawar," Journal of the Royal Society of Arts, LXI (28 February 1913), 400.

[22] Examples of the frequency and duration of British direct administration of native states in Saurashtra:

Number of States of Classes 1-4 under Attachment
(32 is possible)

Year	Number	Source of Data
1873	5 }	Kathiawad Political Agency
1874	6 }	Annual Administrative Report
1896	9	D. H. Kadaka, comp., Kathiawad Directory (Rajkot, 1896), III, 90-94.
1916	4 }	
1917	5 }	
1919	7 }	Kathiawad Political Agency
1920	5 }	Annual Administrative Report
1921	6 }	
1924	7 }	

Agency administration could last for many years. In 1920, for example, four major states were restored to their rulers: Junagadh after 9 years of agency administration; Palitana after 14; Porbandar after 11; and Wadhwan after 3 (K. P. A. Annual Report). Of those states below the fourth class, a vast number were administered by the Agency, usually for reason of indebtedness: 93 in 1896; 142 in 1920; 140 in 1921 (sources, ibid.).

tion; dispensaries run by Vaid specialists in Indian forms of medicine; a central library and museum; and a hospital for women. To finance the central institutions, an annual general fund of about Rs. 121,000 was assessed from the 32 states of the first four classes in proportion to their revenues.[23] To further tighten control without expending their own resources, the British began to pressure the states into bureaucratizing their make-shift processes of rule. To evaluate the change, rulers of each of the larger states were expected to compile annual administrative reports on the various departments of state administration.

These shifts toward centralization, bureaucratization, and the creation of "modern" institutions formed the crux of the British attempt to improve the princely states while retaining control and low expenditure. These policies vitally affected the urban system: First they called for an enclave pattern of development in individual urban areas rather than for spreading the new institutions broadly across the peninsula. Development would be fostered, but in general restricted to the expanding British political capital in a Civil Station under direct British jurisdiction at Rajkot and in another Civil Station at Wadhwan which, as the land gateway to Saurashtra from the mainland, served as a commercial hub. All of the new institutions noted above were headquartered in Rajkot.

The British enclaves became a world apart. The greater political liberality and judicial even-handedness of British administration, as compared with that in most of the native states, made the Stations desirable residences for people concerned with civil liberties. After 1919, when the nationalist movement shifted to mass organization, it focussed on the Rajkot Civil Station as a central location where British law obtained. British law, whatever its restrictions, was generally more benign than that of the states. In addition the transportation and communication facilities of Rajkot and Wadhwan were far superior to those elsewhere in the peninsula. And the kinds of professional and business people resident and working in the Stations were those most likely to be politically alert. The British Stations provided the nodes for nationalist organization just as they did for British administration.[24]

---

[23] A Manual of Karbharis' Meetings of Kathiawar States (1870 to 1940) (Rajkot: Under orders of Karbharis' Meeting, n.d.). This volume provides the best administrative account of the central institutions and the interactions of the states in maintaining them.

[24] Cf. The Times of India's rationale for the location of the protest which grew into the 1938-39 satyagraha in Rajkot, 19 October 1938.

Direct British intervention, however, had little effect on the general growth of urbanization in Saurashtra; the proportion of the population living in urban areas grew slowly. Census data show:

TABLE 2

URBAN POPULATION OF SAURASHTRA:
1872-1941

Year	Percentage of Population Resident in Urban Areas of Saurashtra State Region
1872	19
1881	20
1891	20
1901	30
1911	25
1921	26
1931	27
1941	30

The jump from 20 per cent to 30 per cent registered in the decade 1891-1901 may have been an error of the Census; in any case, the plagues of 1901-3 reduced the percentage substantially and the urban population did not recover for decades to come. The British demonstrated possibilities of new styles and qualities in urban life, and the Civil Stations at Rajkot and Wadhwan grew rapidly, but the overall percentage of urbanization did not increase rapidly.

Second, British rulers called for a new transportation network, but it was designed primarily to focus on the central administrative capital at Rajkot and on the land-port city of Wadhwan rather than to connect smaller trade centers in a fine mesh. The new road network was designed primarily to maximize efficient administration and only secondarily to improve trade.

Third, the princes were elevated to a position as a class of rulers--rather than simply landholders--and were cajoled into attending special training programs to learn to govern as Britishers thought they should. A major effect of this policy was to separate the princes from the lower branches of their lineage as well as from the peasantry.

The British program reduced the responsiveness of the town-based landed groups to the demands of the countryside. The towns remained as the integrating links between the national government and the rural peasantry through the

BRITISH ROADS, 1865-1880

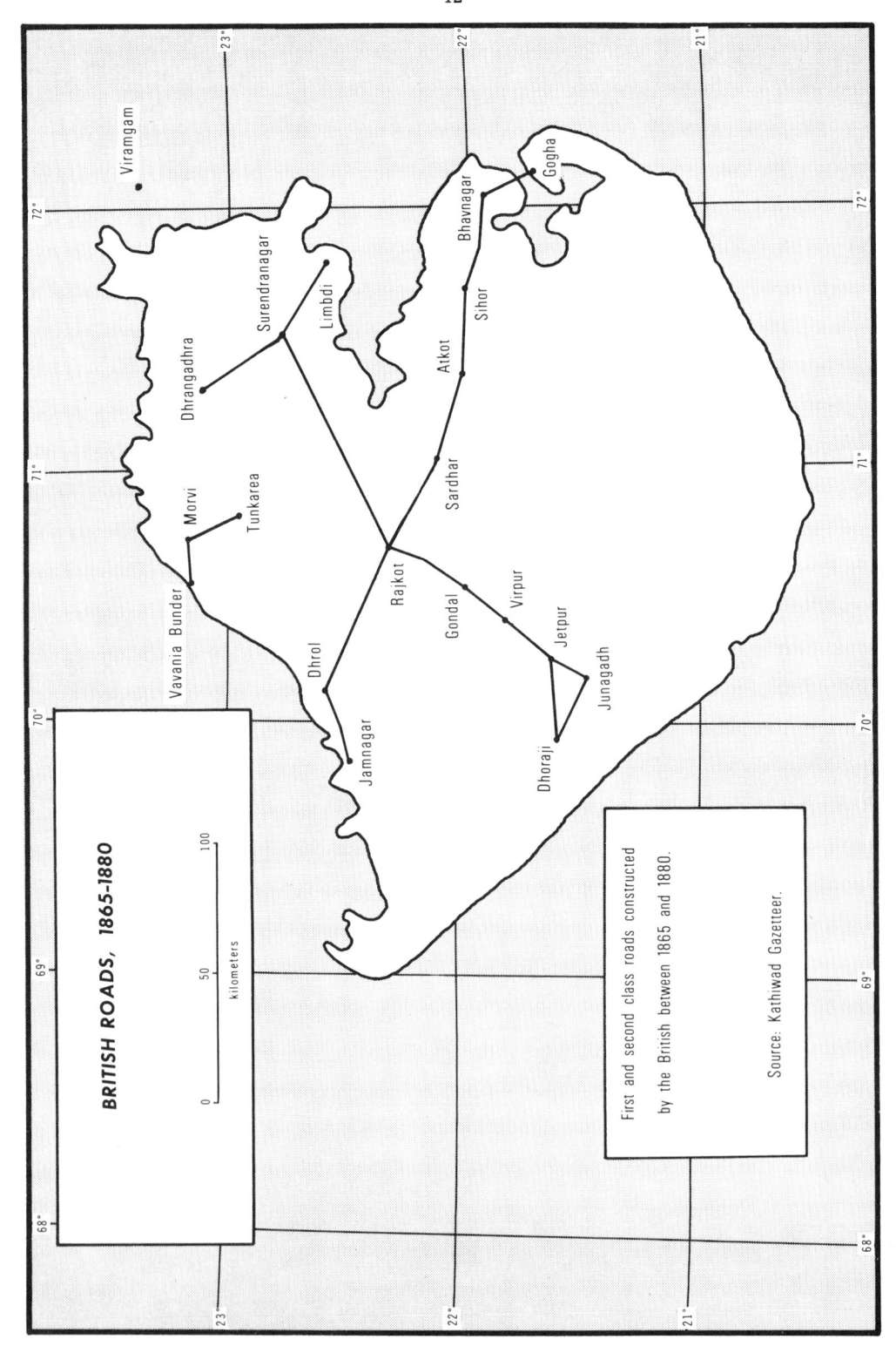

First and second class roads constructed by the British between 1865 and 1880.

Source: Kathiwad Gazetteer.

link of the dominant Rajput lineage elite,[25] but the link was weakened. By stifling any possibility of revolt, the British removed the traditional means of rural protest. The British court system which replaced it would consider only cases among princes. <u>Girasias</u> and others who might have claims against a prince had to go to a court under control of the prince himself. The princes increasingly separated themselves from the lower levels of their lineage and from the peasantry.

The princes were also protected in part from the <u>banias</u>. If the princes fell into arrears in revenue payments, the government took their lands under its supervision until the debts were paid. This limited the former economic power of the moneylenders over the chiefs. Moreover, following the great famine of 1899-1900, the British themselves began to act as moneylenders to all the rulers except those of the highest two classes. They charged interest rates of only 4 to 5 per cent.[26] In addition, again following the 1899-1900 famine, the British promulgated the "life interest principle" which limited debts incurred by princes to their lifetimes only, thus severely limiting the scope of the moneylenders.[27] Within the urban areas, however, the non-Rajput, non-landholding groups could flourish. Peace, administrative requirements of the government, and trade afforded them opportunities. In the larger administrative and commercial centers these urban groups began to challenge the power equation between themselves and the Rajput princes.

From about 1920 throughout India political initiative largely passed from the hands of the British to the rising Indian National Congress movement. Many changes in policy resulted. In Saurashtra these included relaxation of the British system of tariff regulation between Saurashtra and British India (1917); increasing pressure on the princes to enforce personal economies, administrative proprieties, cooperative agreements among states, and to extend greater rights to their subjects; and an administrative redistricting. We largely skip over this period for two reasons. First the major changes came not from British initiatives but from the nationalists. Second, the subsequent assumption of

---

[25] See Richard G. Fox's general argument on this point in his "Rajput 'Clans' and Rurban Settlements in Northern India," in <u>Urban India: Society, Space, and Image</u>, ed. Fox (Durham, N.C.: Duke University Program in Comparative Studies on Southern Asia, 1970), pp. 167-85.

[26] NAI, WISA, 1919, FAM/2 and NAI, WISA, 1918, R/5.

[27] District Record Office (DRO), Rajkot, 1900-01/36.

power by the nationalists effectively subsumed the major reforms and went far beyond them. By focussing on the 1863-1920 period we gain a picture of the transformation of the urban system of Saurashtra at the apogee of British power.

Princely Policies

The princes learned to emulate British models. They were inducted into English modes of life. In 1883, the young princes of Morvi and Gondal travelled to England under British supervision, the first of a long line of Raj Kumar College graduates to make the chaperoned tour.[28] The princes gave donations, too, both to English and Bombay charities and public institutions, even at times when their own states were bankrupt.[29] By 1900 the Kathiawar Times protested that the princes travelled in Europe but not in their home districts; when in their home states they rarely left their showpiece capitals.[30] Perhaps the most English-oriented of the Kathiawar rulers, Ranjitsinhji of Jamnagar, became a world famous cricket player and built a country estate in Ireland and a home in suburban London.[31]

The British controlled and molded the princes through several institutions. One was the college. Another was the ranking system among the states introduced by Keatinge. It allowed various degrees of internal control and responsibility, and provided a framework for reward by elevation and punishment by demotion.[32] As the British became increasingly the community of reference among the chiefs, the ranking became increasingly significant. The British were often called on to adjudicate claims among the princes, most markedly during the lifetime of the special Rajasthanik Court, 1873-1890. Much money and control of land could hinge on the decisions reached. The ultimate weapon of the British was deposition. I have found only one instance of its use in Kathiawad; nevertheless, its threat was real.[33] The British quite effectively made

---

[28] H. H. Maharaja Bhagvatsinhji, Journal of a Trip to England in 1883 (Bombay: Education Society Press, 1886).

[29] Kathiawad Times, 3 November 1890.

[30] Ibid., 6 January 1900.

[31] Roland Wild, The Biography of Colonel His Highness Shri Sir Ranjitsinhji (London: Rich and Cowan, Ltd., 1934).

[32] Wilberforce-Bell, Kathiawad, p. 221; Lee-Warner, Native States, pp. 375-78; NAI, WISA, 1878, Vol. 2, No. 3A.

[33] Kathiawad Times, 12 September 1888.

the princes, if not a class, at least a group apart from the rest of the Kathiawadis, not responsible to their subjects. The links between both the urban landholders and the rural farmers and between the urban rentiers and the other classes within the city were weakened.

The princes began to emulate the urban models set by the British at Rajkot. Even previously they had built elaborate capital cities; now they lavished added attention on them. Capital cities grew at a much faster rate than the other cities of the states, reflecting the enclave philosophy of development. In Bhavnagar State, for example, at the first census in 1872, the capital had 36,000 people while the second largest city held 13,000. By 1941, the last pre-Independence census, the figures were 103,000 to 22,000 respectively. In Jamnagar the relationship in 1872 was 35,000 to 9,000 and became by 1941, 72,000 to 12,000.

Town planning schemes, designed for the capital cities alone, graphically and spectacularly reflect this concentration. Jamnagar, for example, attempted to build a "Paris of Saurashtra." The ruler tore down slums at the center of the city and replaced them with new government offices and private shops. He cut wide roads through densely populated communities, destroying homes and businesses as he went. These new roads were open only to motor traffic; they were closed to the bullock-cart transportation customary in Saurashtra. Spacious new suburban subdivisions were planned. Some were built, but most remain until today as unexecuted blueprints in the Jamnagar Municipality Planning Office. Even on the day-to-day administration of their capitals, rulers spent disproportionate amounts as compared with expenditures for other cities. Compare the data available for Bhavnagar and Junagadh States as shown in the accompanying table. The construction of such grandiose enclaves drew much opposition from professionals and business people.[34]

The princes, propped up by the British and ensconced in their capitals, could afford to pamper themselves and neglect the remainder of the population. Many of the groups with whom they had had to negotiate in pre-British times were now subordinated to them. The British had disbanded the mercenary armies and checked the threats of armed outlawry and revolt from within the state. So, too, British power ended the threat of attack from enemy rulers.[35] Protected from enemies within and without, the princes were less vulnerable to

---

[34] Nawanagar State and Its Critics (Bombay: Times of India Press, 1929).

[35] Wilberforce-Bell, Kathiawad.

TABLE 3

MUNICIPAL EXPENDITURES

Year	Rs.	No.	Rs.
	Bhavnagar City	Other Cities in Bhavnagar State	
1909-10	65,117	9	28,266
1914-15	71,970	9	31,228
1919-20	77,574	9	34,585
1924-25	112,672	9	46,587
1929-30	114,301	9	62,236
1934-35	225,983	9	98,537
1939-40	153,842	9	75,050
1943-44	184,989	9	107,226
	Junagadh City	Other Cities in Junagadh State	
1908-09	21,896	15	10,589
1913-14	41,422	17	25,568
1918-19	56,129	17	42,641
1923-24	69,897	17	66,033
1927-28	72,349	17	65,336

pressure from their subjects. Indeed, it was this freedom that allowed them to attend school, travel, and live as rentiers.

Meanwhile some aspects of British policy, and of princely policy as well, were strengthening the merchant and professional classes. The British presence strikingly increased the opportunities in commerce, especially in cotton export. Note the rapid expansion in the sea trade of the peninsula (Table 4).

The British introduced standard weights, measures, and currency in Saurashtra. Several states had operated their own mints and the British now closed them: Bhavnagar's in 1839, Jamnagar's in 1901, and Junagadh's in 1911.[36] Saurashtra accepted the common rupee currency of British India.

New transportation and communication systems were constructed. The British had the princes finance a trunk-road system under British jurisdiction and administration. In addition, the princes themselves undertook the construction of a railway system. (See accompanying Table 5 and map.)

By removing the Viramgam Customs Cordon in 1917, the British allowed

---

[36] Watson, Statistical Account of Bhavnagar, and Annual Administration Reports for Jamnagar and Junagadh.

TABLE 4

TOTAL SEA TRADE IN RUPEE VALUE

Years	Port of Bhavnagar City Only	Jamnagar State Ports	Junagadh State Ports
1799-1800	3,590,884		
1844		546,000	
1845-1846	363,436[a]		
1864		5,674,725	
1868			7,678,349
1869			6,619,026
1870			8,697,579
1871			7,051,333
1872			7,717,662
1873-1874	17,189,664	6,174,381	
1874-1875	15,986,110		
1875-1876	16,405,822		
1876-1877	16,820,009		
1878-1879	12,674,438	3,568,894	
1879-1880	16,732,103	3,908,340	
1880-1881	18,419,452	4,144,657	
1881-1882	21,122,231	3,941,269	
1882-1883		4,044,522	

Years	All Bhavnagar State Ports	Bhavnagar City Ports	Jamnagar State Ports	Junagadh State Ports
1909-1910	27,208,040	22,201,718	6,325,131	11,084,684
1914-1915	38,870,662	29,811,384	9,599,470	8,716,351
1919-1920	33,009,672	24,519,352	11,952,591	20,383,059
1924-1925	53,297,849	44,642,815	15,212,201	28,766,027
1929-1930	41,690,329	35,839,662	37,775,976	29,903,490
1934-1935	74,466,765	69,346,831	22,186,080	
1939-1940	33,097,629	27,015,540	38,319,873	
1943-1944	37,745,912	32,088,846		

[a]This sharp decline appears to be a result of sharply increased competition from the neighboring British port of Gogha. Cf. Ahmedabad Gazetteer, p. 343.

Saurashtra ports to offer tariff concessions which undercut the cost of importing at British ports. During the years when this was possible, 1917-27, the major coastal states invested large sums in port development: Jamnagar, which heretofore had few port facilities, spent Rs. 16.6 million; Junagadh, Rs. 5.9 million; Bhavnagar, Rs. 4 million. Much of the increase in trade in the period following these investments is reflected in the figures given above.

After 1921 especially, the princes also began to encourage industrial development. In 1921 the three major states combined had fewer than 4,500 work-

TABLE 5

MILES OF RAILWAY AND TOTAL RAIL FREIGHT TONNAGE

Year	Number of Miles of Railway[a]
1880	105
1890	426
1900	548
1910	584
1920	843
1930	1,006
1940	1,084

	Total Freight Tonnage Carried by Rail[b]
1888	109,309
1890	128,966
1910-11	136,133
1914-15	678,300
1917-18	799,765
1919-20	922,193
1920-21	896,724
1921-22	928,465

[a]Sources: A. B. Trivedi, Kathiawar Economics (Bombay: N.P., 1943), pp. 192-95 for 1880-1936; C. N. Vakil, D. T. Lakdawala, and M. B. Desai, Economic Survey of Saurashtra (Bombay: School of Economics and Sociology, University of Bombay, 1953), p. 318, for 1940.

[b]Sources: KPA Annual Administrative Reports for 1918-1922; NAI, WISA, 1890, Vol. XLIII, No. 133, Railways, for 1888-1890.

ers in industry; the accompanying table indicates the growth of industry by Independence. Bhavnagar City alone had some 10,000 industrial workers. The three states together had about 21,000. Opportunities for business and professional people expanded with the economy. The entrance of the princes into industrialization might suggest a <u>rapprochement</u> with the business community, but actually it often increased the friction because the princes usually supported development through restrictive monopolies.

RAILWAYS IN SAURASTRA, 1880-1900

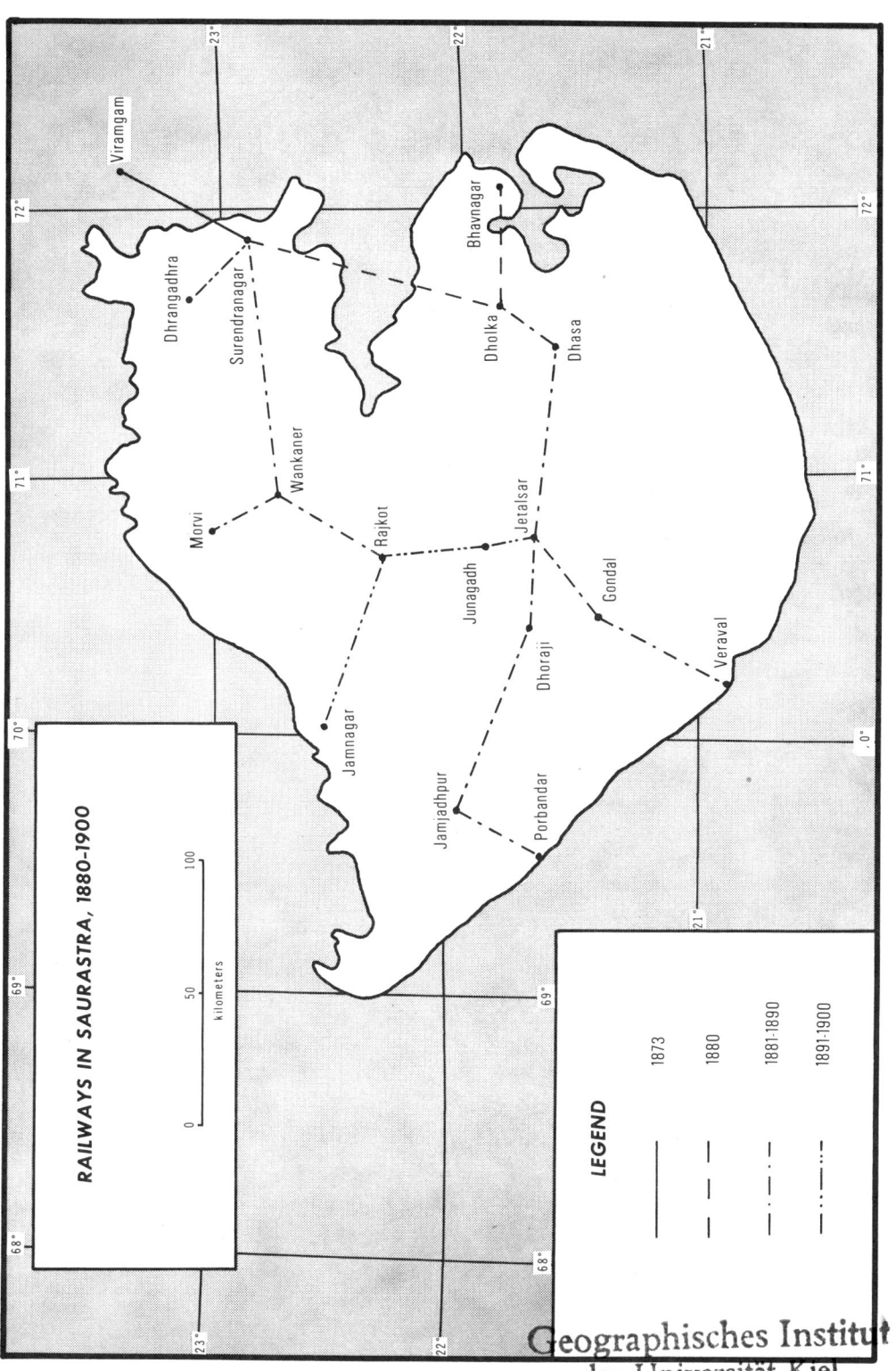

LEGEND

————	1873
– – –	1880
–·–·–	1881-1890
–··–··–	1891-1900

Geographisches Institut
der Universität Kiel
Neue Universität

TABLE 6

NUMBER OF INDUSTRIES IN CITIES

I. Based on Economic Survey of Saurashtra, 1949

City	Number of Factories	Number of Workers
Bhavnagar	113	10,282
Rajkot	98	2,673
Jamnagar	98	5,514
Gondal	NA	NA
Junagadh	NA	NA

II. Based on Government of Saurashtra, Industries of Saurashtra, 1950

City	Number of Workers
Bhavnagar	9,500
Vartej	500
Dhola	500
Botad	500
Jamnagar	6,500
Padadhari	500
Porbandar	4,500
Morvi	4,000
Rajkot	2,500
Dhrangadhra	1,000
Gondal	1,000
Dhoraji	500
Junagadh	500
Veraval	1,000
Keshod	1,000
Manavadar	500

Effects of British and Princely Policies

In the historical context of the urban structure of Saurashtra, British policy exacerbated tensions. It did eliminate battlefield warfare, but by offering new economic opportunities it indirectly encouraged the competition among city-states to relocate in the market place. The intensity of economic competition encouraged the persistence of clashing regional fragmentation rather than integration. Further, by simultaneously promoting the interests of both the stationary, landed princes and the mobile, non-landed merchant and professional classes, the British inflamed the tensions between the two groups. The two historic impediments to political and economic integration among the little kingdoms--competition among states and competition among the urban elites within states--persisted.

The tension between the states increased with the mounting profits from the new trade and transport networks. Table 7 shows the growing importance of these new sources of income in the state budgets.

TABLE 7

PERCENTAGE OF REVENUE FROM LAND, CUSTOMS, AND RAILROADS

Year	Junagadh State			
	Land	Customs	Railways	Ports
1908-09	55	7.6	7.5	
1913-14	47	3.1	17	2.2
1918-19	36	2.9	27	5.5
1923-24	34	2.4	25	7.0
1928-29	40	4.3	22	9.2

Year	Jamnagar State			Bhavnagar State		
	Land	Customs	Railways	Land	Customs	Railways
1909-10	73	4.7	9.8	54	6	24
1914-15	70	3	8.3	56	7.4	27
1919-20	76	10	5	22	9.9	51
1924-25	46	29.8	13.5	28	21	32
1929-30	26	61	5.3	22	47	..
1934-35	28	58	4.5	10	57	16
1939-40	16	58	11.2	23	11	25
1943-44	43	17.6	19.8	24	6	43

Gondal State		
Year	Land	Railways
1914-15	41	21
1919-20	36	25

Sources: Annual Administrative Reports.

Some larger states, notably Bhavnagar and Gondal, adopted free trade policies as conducive to their economic interests. But most princes chose to adopt restrictive trade policies. Their reasons were diverse and often contradictory: restrictive trade policies forced the use of state-owned transport systems; protected infant industries; increased tariff rates; maintained the importance and profits of local city markets instead of allowing subjects to trade more freely elsewhere. Two additional rationales seemed common to all: The traditional responsibility of the princes to insure a food supply at home often inspired export restrictions. More significant even than the economic rationales were the princes' fears that integration of their territories into wider trade networks would diminish their local political authority. Integration into larger co-operating units would highlight the relatively small size and minor importance of any one single state.[37]

Princes imposed customs barriers under one rubric or another.[38] They sabotaged the trunk-road system by allowing bridges and roads to remain washed out and in general disrepair after the monsoon. They diverted commerce from the roads to the railways which they owned. The railway system, begun in 1880 in a spirit of harmony, broke up in acrimonious dispute when the profits proved worth fighting over. In 1911, the multi-state syndicate which controlled the Bhavnagar-Gondal-Junagadh-Porbandar Railway system broke apart into six separate units; the states could not agree on the most appropriate division of profits nor compromise their competitive interests in railway routings.[39] The bus transportation system which promised to develop in the 1920's was crippled by princely restrictions. "Each state had its own separate road transport system unconnected with those of other states."[40]

Most city-states seemed willing to act in protectionist ways against the others; the smaller states seemed most eager. They feared economic absorption. The ruler of Kotda Sangani, for example, feared that if he allowed free trade, his capital city would decline; his subjects would shift their trade to Gondal, eight miles to the west, or to Rajkot, fifteen miles to the north. Like most rulers of the smaller states of Saurashtra, he erected barriers to safeguard the importance of his home marketplace.[41] The presence of multiple

---

[37] Manual of Karbharis' Meetings, p. 439.

[38] Officially, inter-state customs were outlawed by the British. Persistent reports show that the British policy was not successfully implemented. At the time of Independence, inter-state duties collected by all States combined were valued at Rs. 4,500,000. Government of Saurashtra, Memorandum Presented by Government of Saurashtra to the Part B States (Special Assistance) Enquiry Committee, June, 1953, p. 21.

[39] The record of the dissolution of the BGJP railway syndicate unfolds in NAI, WISA, 1911, RY/2 and RY/3, "The Breakup of the BGJP Railway."

[40] C. N. Vakil, D. T. Lakdawala, and M. B. Desai, Economic Survey of Saurashtra (Bombay: School of Economics and Sociology, University of Bombay, 1953), p. 323.

[41] Subsequent events suggest that the ruler's fears were justified. After

jurisdictions and headquarters towns in the peninsula multiplied the leverage of the smaller city-states at the expense of the larger. British policy conserved these small units and protected them from change.

By strengthening both the princes and the non-landed urban classes at the same time, the British increased the tensions within the capital cities as well as between them. As the opportunities for trade increased, pleasing the merchants, the princes imposed increasing restrictions on trade. The blockages in the inter-state transport system stung the merchants. The development of industries as state monopolies or as state supported monopolies, angered them still more. At first the monopolies boosted industrialization, but after a time they appeared restrictive. By the time of the Rajkot Satyagraha of 1939, for example, opposition to the prince of the Rajkot city-state included opposition to his state monopolies: the state bank, the state textile mill, and the state powerhouse. Merchants, in particular, withdrew their deposits from the bank, severed their electrical connections, and refused to trade the mill's products.[42]

The pettiness of state rulers frustrated the aspirations of promising professional and business people and drove them from the peninsula. Mahatma Gandhi's experience provides an early example. A native of Saurashtra, Gandhi attempted to found a legal practice in Rajkot in 1892 on his return from legal training in England. After a few months he left in despair, lamenting that "the intriguing atmosphere of Kathiawad was choking."[43]

Over the years the tempo of emigration quickened. In 1911 the census reported that 6 per cent of the total Saurashtra-born population was resident in other parts of India; by 1931 the figure had risen to 10 per cent. More significantly, the emigrants were of the traditionally mobile high castes, particularly merchant castes, in the prime of their working years and occupied in business and the professions. Breakdown of the 1911 Census data on emigration of Saurashtrians to Bombay City, the main destination, showed almost half of the emigrants as belonging to trader castes and 11 per cent more as Brahmins. On ar-

---

Independence and merger in 1947-58, the population of Kotda Sangani town dropped from 4,219 in 1951 to 4,194 in 1961 while the population of the district was rising about 30 per cent. The town was declassified as urban in the Census of 1961 and reclassified as rural.

[42] Mohandas Karamchand Gandhi, The Indian States' Problem (Ahmedabad: Navajivan Press, 1941).

[43] M. K. Gandhi, Satyagraha in South Africa (Madras: S. Ganesan, 1928), p. 67.

rival at their destination, they continued to pursue careers in business and the professions. (See Table 8.) Further, Saurashtra showed only 28.9 per cent of its population in the prime working age of 20-40 while Bombay Presidency as a whole showed 32.9 per cent in 1931.

Emigration had always been a strategy open to the non-landed urbanites in Saurashtra at times of discontent, and the princes recognized its significance. The annual administration report for Jamnagar State noted in 1924-25: "Unless we expand our trade and commerce, and widen the existing channels of our industry, apart from agriculture, the more enterprising members of our population will continue deserting us."

The emigrants were disaffected from both the princes and the British who preserved them. The Census of 1931 remarked on "the well known fact that the civil disobedience movement in several areas of the Bombay Presidency, and especially in Bombay City itself, was greatly favored by the trading classes whose original home is in Kathiawar and Kutch."[44] These emigrant Saurashtrians supported and read Saurashtra, the major communication organ of the nationalist movement in the peninsula. The Gujarati language weekly was published just across the Saurashtra border in Ranpur, in British India, from 1921. Its editor, Manilal Kothari, evidently found this a more congenial home than was princely Kathiawad. Indeed, many states banned the newspaper. Saurashtra, in fact, sold most in Bombay among the emigrant Saurashtrians there. Along with its post-1932 successors, Fulchhaab in the peninsula and Janmabhoomi in Bombay, the newspaper tied the peninsula to the outside world and provided coverage which made Saurashtrians begin to see their common problems and common condition as requiring revision of the princely system and the overthrow of the British who propped it up.

The founding of a Kathiawar Chamber of Commerce in 1900 to protest the unfair competition of the merchants of British India and to advocate Saurashtra's interests further reflects the degree to which Saurashtrians began to organize even across state lines. Modern transport and communication was forging from the fragmented peninsula a geographically unified political organization.

The struggle between the landed rulers in command of the cities and the non-landed urbanites came to a head by 1938-39 in the Rajkot Satyagraha. John Wood (in a recent Ph.D. dissertation) described the conflict:

> Of lasting importance for the politics of Saurashtra was the fact that the Rajkot Satyagraha pitted not only the Gandhians against the prince, but also the

---

[44] Census of India, 1931, Vol. X, Western India States Agency, Part I, p. 21.

TABLE 8

EMIGRATION FROM KATHIAWAD: 1911

I. Total emigrants--171,582, or 6 per cent of total population born in Saurashtra

II. Emigrants to the major cities of Western India

City	Number of Emigrants	Percentage of Total Emigrants from Saurashtra
Bombay	58,775	34
Karachi	14,728	9
Ahmedabad	8,335	5
Surat	3,394	2
Total of 4 cities	85,232	50

III. Partial caste distribution of male emigrants to Bombay, 1911. (Total male emigrants numbered 38,586, or 64 per cent of total emigrants)

Caste of Emigrants	Number	Percentage of Total Male Emigrants
A. Members of traditional trading communities	17,351	46
1. Hindu Vanias	6,344	
2. Jain Vanias	2,388	
3. Memons	2,365	
4. Lohanas	2,023	
5. Khodas	1,556	
6. Bohoras	1,415	
7. Bhatias	1,260	
B. Total Brahmins	4,152	11

IV. Partial distribution of male emigrants working in Bombay by occupation, 1911

Occupation	Number of Male Workers	Percentage of Total
Total male workers	30,839 (80 per cent of total males)	100
Shopkeepers	10,940	35
Artisans	5,630	18
Clerks	2,675	9
Domestic servants	2,716	9
Laborers in textile mills	388	1
Other "laborers"	859	3

Source for all units of table: Census of India 1911, Vol. VIII: Bombay, Part II, pp. 199 and 215.

non-violent urban castes against the martial and depressed castes of the rural areas [and their urban leadership cadres]. The leadership of the struggle was entirely in the hands of Brahmins and Banias.[45]

Among the half-dozen top leaders in the pre-arranged chain of command were four banias and two Brahmins.[46]

The lines of class conflict within the city were clearly drawn. As the struggle intensified, the non-landed business and professional classes attempted to swing the balance in their own favor by involving the rural population in political life. For the first time during the nationalist campaign it appeared that the rural agricultural population might be brought into the political and market processes. Rural groups had, heretofore, been excluded from participation in the political and economic life of the princely headquarters towns. As noted, British power precluded rural uprisings, the traditional form of revolt. This left the only systemic means of political activity to be direct petition to the prince. Moreover, throughout India the British feared the political consequences of rural land passing into the hands of urban moneylenders.[47] Therefore in Saurashtra, as elsewhere, the British forbade city businessmen from buying agricultural land, thus severing the rural land economy from the general marketplace. Agriculture in much of the peninsula was encouraged not to orient toward the market; although in the large and more progressive states revenues were collected in cash as early as the 1870's (Bhavnagar), many of the smaller states continued collecting in kind until Independence. The desire to ensure an adequate internal food supply without chancing market involvement may be a valid explanation for a small state discouraging market agriculture, but it nevertheless served to create a dual economy. The perpetuation of veth, or unpaid corvée labor, commonly exacted in the Saurashtra countryside, further indicated the lack of penetration of the money economy into rural areas.

The perpetuation of tenancy at will through the withholding even of occupancy rights--much less rights of alienating land--also tended to keep the rural economy encapsulated from the urban. At the commencement of British rule,

---

[45] John R. Wood, "The Political Integration of British and Princely Gujarat: The Historical-Political Variable in Indian State Politics" (unpublished Ph.D. dissertation, Columbia University, Faculty of Political Science, 1971), p. 147. Wood is correct in seeing the martial and depressed castes as largely rural in composition, but their leaders were the urban-based Rajput rulers.

[46] Ibid.

[47] Walter Neale, Economic Change in Rural India (New Haven: Yale University Press, 1962) has a most lucid discussion of this issue.

even up to the 1880's, all khalsa land was owned by the state; cultivators had no rights of alienating land, nor of mortgaging it for loans, nor of maintaining ownership of improvements. The first state to grant rights of permanent tenure and of alienation of agricultural land was Gondal, acting in 1893. Gondal added that while land could be sold at the will of the occupier, it could not be sold by a court in order to pay off debts.[48] This protected the landholder from the moneylender. Bhavnagar, which granted occupancy rights in 1872, did not allow the alienation or mortgage of land, though subletting was allowed in certain circumstances.[49] Junagadh granted occupancy rights while under minority administration, 1911-20;[50] Jamnagar granted them during the administration of Ranjitsinhji.[51] Only about ten states granted even such permanent occupancy rights; Rajkot, like the vast majority of smaller states, never did.[52]

As the anti-princely forces went to the countryside to organize the peasantry in opposition to the ruler, the princes responded in two ways. Some of them, some of the time, granted concessions to the peasants in the form of permanent occupancy rights, allowing the alienability of land, and enforcing laws against corvée labor. An alternate response, however, was the attempt to seal off the countryside from the urban organizers. In many states the urban classes were permitted to propagandize within the capitals, but were forbidden to carry their message to the countryside. Most dramatically, the Limbdi Satyagraha of 1939 resulted in violence and mass flight as the organizers refused to heed the prince's restrictions against rural organizing and were met with retaliation by armed force. Again the contest between banias and the ruler was clear.

In February 1939, as agitation in Limbdi began, organizations spawned by the ruling family and staffed by officers and servants of the state issued leaflets

---

[48] Nihal Singh, Shree Bhagvat Sinhjee: The Maker of Modern Gondal (Gondal, 1934), p. 181.

[49] Javerilal Umiashankar Yajnik, Gaorishankar Udayashanker, C.S.I. (Bombay: Education Society's Press, 1890?), p. 59 and Bhavnagar State, Some Important Papers Relating to the Revision Settlement of the Mahals of the Northern Division, p. 7.

[50] Annual Administration Reports of Junagadh State.

[51] John da la Valette, An Atlas of the Progress in Nawanagar State (London: Mssrs. East and West Ltd., 1931?).

[52] Government of Saurashtra, Memorandum to Part B States Enquiry Committee, p. 38.

stating that the Praja Mandal, the chief local organization of nationalist protest, "was really only a Mandal of the Banias, and did not enjoy the support of the majority of the people of the State."[53] State officials

> compelled the Banias living in the villages to leave the State. The Bania servants of the State were dismissed, pensions were stopped, while the property of those who had left the city began to be looted in a systematic manner. . . . In order to force out the few merchants who had remained behind in the village of Panshina, the railor, the potter, the barber, the shoe maker and other artisans were told that they should not do any work for the merchants.[54]

Gandhi himself, the <u>bania</u> leader of the nationalist movement, wrote of Limbdi,

> The heaviest blow has been aimed at the hated Bania, who was at one time the State's friend, favorite, and main supporter. But he was to be crushed because he dared to think and talk of responsible government, dared to go amongst the peasantry and tell them what was due to them and how they could get it.[55]

In trying to organize the countryside, the urban professional and business classes were attempting to spread their rising aspirations and ideologies to a rural peasantry not yet so optimistic. The non-landed urbanites pressed for a redistribution of political and economic power, and for a new ethos of regional development in Saurashtra. They attempted to extend political activity both vertically downward by involving the peasantry, and horizontally in space by forging peninsula-wide political organizations.

Meanwhile, most princes thought that adequate progress was already being made. By building up their capital cities; by substantial investment in ports, railroads, and industries; and by administrative reform, the princes were moving from their earlier policy of the state as agricultural fief to a new, as yet undefined policy. And, in a process which is often overlooked, they did in fact create substantial urban, transport, and communication infrastructures which would serve the peninsula well after Independence and the merger of the jealous city-states. But, as the Bhavnagar People's Assembly pointed out in 1928, "no matter how fast improvements come, it is but natural that the people want more."[56] In the light of the demands being made, the princes moved too slowly. The expectations of the urban professional and business classes were rising

---

[53] Narhari D. Parikh, <u>Sardar Vallabhbhai Patel</u> (Ahmedabad: Navajivan Publishing House, 1956), II, 379.

[54] Ibid., p. 382.

[55] Ibid., p. 383.

[56] Bhavnagar Prajaa Parishad, <u>Trijun Adhiveshan</u> (Botad, 1928).

quickly. They sought faster progress in a more open atmosphere, either through emigration or political revolution. In the years immediately preceding Independence both of the antagonisms in the urban system had been rubbed raw. Mobile urbanites and landed princes confronted each other directly within the capital cities, and agitation for regional, horizontal integration directly opposed princely desires for territorial fragmentation into personal fiefs. (For an extreme picture of one ruler's view of an appropriate political system for his small state, see Appendix 2, Lakhaji Raj's plan for Rajkot state.)

Summary and Conclusions

British rule heightened the two tensions inherent in the indigenous urban system. By protecting the princes and their domains, the British transferred the locus of inter-city competition from the battlefield to the marketplace. As opportunities for trade increased, competition took jealously isolationist forms in most of the states. In addition, British policies which strengthened both the merchant and professional classes on the one hand, and the princes on the other, increased the traditional conflict between these two urban groups. The princes saw that continued political control of their states required a substantial degree of encapsulation and isolation, while the professionals and the long-distance traders favored increased external contacts. By promoting a limited degree of economic modernization, British and princely policies first created, and then frustrated, a revolution of rising expectations.

Several caveats must hedge this argument. The generalizations here clearly do not fit all the city-states even within an area so small as Saurashtra. I have elsewhere[57] examined five city-states in some detail and have found obvious differences in policy. In addition, the categories of princes, professionals, and merchants are not adequately specific. At present I can only cite illustrative examples of each group and of their interactions, but I lack the data to reconstruct them comprehensively. The generalizations, however, do explain most of the cases and they do account for both the great territorial fragmentation and the incipient class conflict within the cities of the peninsula.

---

[57] Howard Spodek, "Urban-Rural Balance in Regional Development: A Case Study of Saurashtra, India, 1800-1960" (unpublished Ph.D. dissertation, University of Chicago, 1972).

CHAPTER III

SAURASHTRA SINCE INDEPENDENCE: THE ACHIEVEMENT OF AN URBAN-RURAL REGIONAL BALANCE

> Can anyone consider with equanimity the picture of India technologically advanced in the urban sector and static in the rural sector?
> --U. N. Dhebar, Chief Minister of Saurashtra State

Introduction

Independence in 1947 brought an end to 150 years of British rule. The departure of their British supporters paved the way for the fall of the princes and the subsequent merger of their separate states into one political unit, the United States of Saurashtra. An organized and coherent group of nationalist politicians came to power and set about a restructuring of economic status. They chose Rajkot as the capital of the new state, but all the other state capitals lost their historic political functions.

Independence undercut the balance of power among urban elites. Ruling princes whose power had been based on rural holdings and backed by the British lost out as the Brahmin and Bania professional and business classes came to dominate the region's politics. In many areas of India similar changes were occurring, but perhaps in no other was the transition so dramatic. The Congress Party, the political home of the urban business and professional classes, carried the first general election to the Saurashtra State Legislature in 1952 by the most decisive majority in all of India, with 53 seats of a total 57.[1] The leading members of the cabinet during the eight years of the existence of the Saurashtra State were: U. N. Dhebar, a Nagar from Jamnagar who carried on his political work in Rajkot; Rasiklal Parikh, a Bania from Surendranagar Dis-

---

[1] Wood, "The Political Integration," p. 171.

trict; Manubhai Shah, a Bania who had been working with the Delhi Cloth Mills in Delhi; Balvantrai Mehta, a Bania from Bhavnagar; and Ratubhai Adani, a Bania of Junagadh. Of the other cabinet members who served at different times, all were city people, all professionals and businessmen, and all Banias and Brahmins. Together they represented the major geographical areas of the peninsula.

From long years of opposition to the princes and of grass-roots organizing in the countryside, the Congress leaders had committed themselves to redressing the balance of power and wealth between the urban-based princes and the rural peasantry. They were not, however, anti-urban. They sympathized, too, with the poorer classes in the city, notably the artisans and the workers. Also, as businessmen and professionals they understood the significance of urban centers to general regional development. They acted to transform the cities into productive centers which could work in tandem with an increasingly productive countryside. Their assumption of political rule set the stage for massive change in the economic life of Saurashtra, and significant, but lesser, change in the social patterns.

The political merger of the former princely states gave impetus also to major change through unification of transportation and trade networks and through a new and wider focus for planned development. Several years before Independence, the ruler of Jamnagar had proposed a plan for a loose union between not only the separate states of Saurashtra, but including those of Rajasthan as well. The "Jam Jyooth Yojna," however, was rejected by the jealous princes.[2] The British enforced an "Attachment Scheme" after 1944 which brought the smallest states under the administrative control of the largest in Saurashtra.[3] But so long as the princes held to their prerogatives, effective cooperation was hobbled.[4] Independence and merger brought both internal unifi-

---

[2] Ibid., Sec. V, pp. 4-5.

[3] See Government of India, Ministry of States, White Paper on Indian States (Delhi, 1950), p. 38 and Wood, Sec. V, pp. 4-5.

[4] Cf. Janmabhoomi (Bombay), August 8, 1945, reporting that in the last years of the Raj a number of plans were designed in Kathiawad especially, either by the Resident or with his willing consent, to bring the states into workable, larger relationships. But the states refused. In Rajkot on August 6, 1945, the prince rejected a scheme for a common policy on road development in Kathiawar, in line with proposals for princely states all over India. They feared that if they gave up one power, slowly all their powers would be eaten away. I am indebted to Kantilal Hathi for calling my attention to this article and several others in his personal file of clippings on events leading to political merger.

cation and considerably closer ties between the peninsula and the rest of India.

Merger made the class conflict between the princely versus the professional and business elites more difficult to pinpoint geographically than earlier conflicts and resolutions had been. The merger of the states into one political unit and the much closer weaving of Saurashtra into the national political fabric reduced the autonomy of the individual cities and rendered the city inappropriate as the focus of study. The Government of India's assistance and overall programs--as for instance in giving Rs. 6 crores of assistance in loans and grants to the Saurashtra government for its first five-year plan of 1951-56--introduced an important extraneous influence. Similarly, the creation of an active, powerful, state-wide government and the end of politically autonomous city-states terminated the significance of the local urban areas for the development of city-state micro-regions. Individual cities, their immediate hinterlands, and indeed the whole of the peninsula were subsumed into much more comprehensive political units.[5] This change in the scale of political units may be the most important change brought to Saurashtra by Independence.

Certainly the new political leaders saw merger as challenging as well as vital. They saw that the enlarged and integrated territorial units for which they had worked now posed psychological as well as political difficulties. An official statement of the new government explained:

> . . . the popular mind today is nervous. They have lost their little world of Princes and Chieftains which they knew for decades. In that little world, there was undoubtedly more misery but certain advantages, too. They were accustomed to see all that belonged to them before their naked eyes. If there were debts to be paid, they knew what they were and what they meant. If there were assets, they knew how they were utilized whether for their betterment or for the individual aggrandizement of the Ruler. It was a feudal framework but it was a paternal framework too in some respects. The dreams of Ram Rajya which they had cherished when the Congress was fighting for their liberation have led to the disappearance of that little world but not to the appearance of a new and better world.[6]

The British departure, too, brought significant readjustments in the economic life of Saurashtra. It removed oppressive economic restrictions particu-

---

[5] A similar difficulty in locating an appropriate scale of focus for urban and urban-rural analyses for the United States is confronted by Daniel Elazar, Cities of the Prairie (New York: Basic Books, 1970) in the political context of federalism, and by Otis Dudley Duncan et al., Metropolis and Region (Baltimore: Johns Hopkins University Press, 1960) in the spatial context of economic development.

[6] Government of Saurashtra, Memorandum Presented . . . to the Part B State . . . June 1953, p. 13.

larly on exports from Saurashtra to British India. For example, the British had imposed a ban on the export of salt from Saurashtra to British India and the explosion of salt exports following the departure of the British was one index perhaps minor, but symbolic in light of Gandhi's use of it in his campaigns, of the economic effects of the political change.

TABLE 9

SALT EXPORTS (TONS)

Year	Tons of Salt
1945-46	5,126
1946-47	10,284
1947-48	7,483
1949-50	234,644
1960-61	389,339

Sources: Economic Survey of Saurashtra, p. 293, for 1949-50; Gujarat State, Ports Traffic Review of 1960-61.

To some extent, also, the flowering of the Saurashtrian regional economy after Independence must also be seen as a maturation of developments already initiated during the princely period. We have already argued that the princes created an urban infrastructure which served as a skeleton on which future generations might build.[7] Several states had carried out model reforms which had increased their productive potential. Gondal had allowed rights of ownership in the land to the peasantry, simplified its tax system, built roads, fostered education, and encouraged cooperatives. Bhavnagar had instituted a land mortgage bank and village panchayats. The new government had successful local models for adaptation throughout the peninsula.

In addition to the man-made advantages, Saurashtra inherited a major ecological benefit, "one of the most favorable man-land ratios . . . in the Indian

---

[7]Cf. the Danish experience of older, more politically restrictive cities becoming centers of entrepreneurial expansion especially in conjunction with the rural economy, after a new group of political leaders gained power after 1870. Johnson, The Organization of Space in Developing Countries, p. 24. Cf. also the transformation of the Japanese castle town into a growth point after the Meiji Restoration; John Whitney Hall, "The Castle Town and Japan's Modern Urbanization," Far Eastern Quarterly, XV, No. 1 (November, 1955), 37-56.

Union."[8] With 343,000 farm families on 8½ million acres, Saurashtra had an average of about 25 acres per family.[9] In most parts of the peninsula, however, the man-water ratio was not favorable, excepting the Gondal region which benefited from the Bhadar River System.

But the main root of the transformation in the post-Independence economy of the Saurashtra region, and in the redressing of the urban-rural balance was the political revolution within the peninsula. This revolution, though peaceful, completely changed the dominant political actors and ethos of the region. It removed the princes from power and in their place brought those other historical urban leaders, the business and professional groups. It replaced the "Bhom Raj," based on aristocratic landlord values, with "Bania Raj," based on merchant values.[10]

I cannot "prove" that this political restructuring was the key to change, but as I present the programs and results of the Government of Saurashtra, I shall draw comparisons to programs in other parts of India which strongly suggest that the politics of Saurashtra was different and decisive. I shall contrast it, in passing, with other areas of India, particularly princely areas, which also saw an end to princely rule and the consolidation of a unified state but which did not advance nearly so quickly.

The Infrastructure

No area of India had been so divided into separate, tiny jurisdictions as Saurashtra had been with its 222 units in the small peninsula of 25,000 square miles. The princes, particularly those who controlled smaller states, had acted to isolate their states economically and to narrow the extent of market facilities to their own borders by rigging tariffs, regulating imports and exports, allowing roads to decay, and attempting to attain internal self-sufficiency in food rather than moving their states into a wider exchange economy. Uni-

---

[8] Joseph E. Schwartzberg, Occupational Structure and Level of Economic Development in India: A Regional Analysis, Monograph Series, No. 4 (New Delhi: Census of India, 1961), p. 165.

[9] Government of India, Central Statistical Organization, Statistical Abstract, India 1956-57 (Delhi, 1958), pp. 498-99. The relevant table is reproduced as Graph 2 in the present chapter of this study, below.

[10] This contrast is based on Tod's statement of Rajput values. James Tod, Annals and Antiquities of Rajasthan (London: George Routledge and Sons, Ltd., 1914), I, 394.

fication of the peninsula in 1948 and government policies for improvement of the transport network enlarged the market, fostering economic rationality.[11] The road system of Saurashtra had been poorly integrated and poorly maintained. The new government acted to supply missing links where necessary and to expand the road network and the facilities for road travel (see Tables 10 and 11).

TABLE 10

ROAD MILEAGE

Year	Miles of Metalled and/or First Class Roads	Total Road Mileage
1865	0	No data
1880	330	550
1950	1,880	3,358
1961	3,085	5,515

Sources: 1865 and 1880 from Kathiawar Gazetteer; 1950 from Economic Survey of Saurashtra; 1961 from Census of India 1961, V, Part I-A(i), 149; for the data on metalled roads, see the District Census Handbooks for each of the six districts of the Saurashtra peninsula for 1961. Data for the years between 1880 and 1950 are unobtainable since most roads were a subject of local state administration and the records, if available at all, are nowhere assembled.

Post offices, schools, and banking offices--the latter often in the form of branches of government administered banks--were expanded rapidly. Electrical power generation was also stepped up several times. Prior to Independence a few states had operated small generators for their capital city, or at least for the public, princely buildings located there, but now the effort to supply power was significantly broadened and the plans for supply organized on a peninsula-wide (and later a still wider regional) basis.

Private individuals and business responded to the new opportunities and the new spirit of expansion. Literacy increased; in rural areas in particular it doubled. Publications multiplied. (Their freedom of expression had been se-

---

[11] For the first few years, years of famine, the new Saurashtra Government also attempted self-sufficiency in food through bonus schemes, but at least the market was broadened to cover the entire peninsula. And after a few years the entire policy of food self-sufficiency was scrapped.

TABLE 11

ROAD TRAVEL FACILITIES, 1948-57

Index	1948[a]	1950-51	1955[b]	1956[b]	1957[b]
No. goods trucks	792	1,987	2,548	NA	NA
No. taxis	400	459	729	NA	NA
No. private cars	NA[c]	3,261	3,728	NA	NA
No. bus route miles/day	NA	6,000	NA	13,726	19,198
No. bus passengers/day	NA	NA	NA	NA	14,803

Sources: 1948 and 1950-51: Economic Survey of Saurashtra, pp. 323-24. 1955: Annual Administrative Report of Saurashtra. 1956 and 1957: Statistical Abstract of Bombay State, 1957-58.

[a] Pre-integration.

[b] Road transport system nationalized.

[c] In 1949-50, 2,970 private cars reported.

TABLE 12

POST OFFICES

Year	No. Post Offices	Source of Data
1872-73	39	Kathiawar Gazetteer, p. 234.
1879-80	73	Kathiawar Gazetteer, p. 234.
1914-15	413	KPA Annual Administration Report.
1920-21	413	KPA Annual Administration Report.
1950-51	385 in four districts: Surendranagar, Bhavnagar, Junagadh, and Amreli. Others N.A.	Based on computations from Census of India 1961, Vol. V, Part I-A(iii), p. 122.
1960-61	1,142	Census of India 1961, Vol. V, Part I-A(iii), all six districts.

cured.) Port traffic increased. Joint stock companies flourished, and most notably factories increased particularly in the rural areas.

Land Policy

Government policy explicitly favored economic growth in the countryside. U. N. Dhebar, the Chief Minister and a committed Gandhian, asked rhetorically,

TABLE 13

SCHOOLS

Year	Number of Schools	Number of Students	Percentage of Population
1875	399	20,576	0.86
1910-11	1,129 + 324 private	79,972 +15,125 private	
1919-20	1,741	129,676	
1938-46	2,210	193,868	6.8
1948	2,124	160,650	
1951	2,614	202,850	
1960-61	5,007 primary 283 secondary 23 higher education	512,059 74,742 7,894	

Sources: 1875 from KPA Annual Administration Report; 1910-11 from Gazetteer of Kathiawar, 1914, p. 44; 1919-20 from KPA Annual Administration Report; 1938-46, 1948, and 1961 from Economic Survey of Saurashtra, pp. 338 and 342; 1961 from Census of India 1961, V, Part I-A(iii), 64-67.

TABLE 14

DISTRIBUTION OF SCHOOLS IN VILLAGE AREAS--1961

District	Total No. of Villages	Villages with Schools	Percentage	Area per School in Sq. Miles	No. of Students on Roll
Amreli	595	509	85.55	4.12	48,517
Rajkot	855	785	91.81	5.17	61,579
Jamnagar	701	619	88.30	6.11	40,734
Junagadh	1,126	886	82.88	3.79	59,927
Bhavnagar	881	817	92.74	4.11	47,603
Surendranagar	661	551	83.36	6.83	27,793
Total	4,819	4,167	86.47	....	286,153

Sources: District Census Handbooks for each district, 1961.

TABLE 15

BANKING OFFICES

Year	Number of Banking Offices Open
1910	2
1921	3
1927	4
1928	6
1932	8
1943	9
1944	13
1945	20
1946	21
1947	27
1948	29 before integration
1948	35 after integration
1949	37
1950	52
1951	61 plus six more for which the year of opening is not available; total 67
1955	123

Source: Economic Survey of Saurashtra, p. 328.

TABLE 16

ELECTRICAL GENERATION

Year	Kilowatt Hours Generated
1948	21,383,000
1950-51	30,948,000
1955	56,600,000
1960-61	153,067,000

Sources: 1948, Industries of Saurashtra, pp. 30-35; 1950-51 and 1960-61, Census of India 1961, Vol. V, Part I-A(iii), p. 117; 1955, Indian Institute of Public Opinion, Quarterly Economic Report, VIII, No. 2 (October, 1961), 33.

TABLE 17

LITERACY

Literacy Rates (percentages) over All Regions Considered

Year	Region Covered	Percentage of Total Population Literate	Percentage of Males Literate	Percentage of Females Literate
1911	Kathiawad	10.1	17.9	2.0
1921	Western India States Agency	9.9	16.5	3.0
1931	Western India States Agency	12.5	20.4	4.3
1941		NA	NA	NA
1951	Saurashtra State	18.5	26.3	10.5
1961	Saurashtra Peninsula and Kutch[a]	27.4	37.8	16.4

Literacy Rates, Urban and Rural, for Population Age 5+ (percentages)

Year	Region Covered	Urban		Rural	
		Male	Female	Male	Female
1951	Saurashtra State	52.9	26.0	20.7	5.4
1961	Rajkot Division[b]	65.6	38.0	35.6	11.9

Sources: Census of India for respective decennials.

[a] Rajkot Division of Gujarat State.

[b] Includes Kutch.

TABLE 18

NEWSPAPERS PUBLISHED[a]

Year	Dailies	Weeklies	Fortnightlies	Monthlies
1950-51	5	8	1	5
1960-61	5	14	8	31

[a] All newspapers published in Gujarati except one monthly in 1950-51 in Hindi, and one in 1960-61 in English.

TABLE 19

PRINTING PRESSES

Year	Number of Printing Presses	Source of Data
1880	9	NAI WISA, Vol. II, No. 3, KPA Annual Administration Report.
1890	18	NAI WISA, 1890, Vol. V, No. 4, KPA Annual Administration Report.
1924	52	NAI WISA, 1924, ADM/1, No. 2, FPA Annual Administration Report.
1961	212	District Census Handbooks, six districts.

"Can anyone consider with equanimity the picture of India technologically advanced in the urban sector and static in the rural sector?"[12]

The new Congress government had come to power in Saurashtra with virtually no debts to landlord interests. Land reforms were quickly legislated and decisively enforced.[13] The Congress package of land reforms was designed to abolish the middleman in agricultural taxation, to channel the agricultural revenues directly from the tiller to the state, and to secure land to the tiller. It also ended the crop-share system and replaced it with a cash system of revenue, bringing the villagers more directly into the market economy. It reduced the total burden of revenue both by a reduction in the land revenues and by the abolition of additional cesses which had plagued the tenants-at-will.

No time was wasted. "A proclamation on the 15th April, 1948, the date

---

[12] U. N. Dhebar, "Role of Khadi in Village Industries," in Nawanagar Chamber of Commerce, Industrial Information Center: Souvenir (Jamnagar: The Chamber, 1965).

[13] R. R. Mishra, Effects of Land Reforms in Saurashtra (Bombay: Vora and Co., 1961) is a useful review of the entire land reform program in Saurashtra, replete with a survey of 64 sample villages to test the results of the legislation in the field. The study was funded by the Research Programs Committee of the Planning Commission, Government of India. I know of no similarly comprehensive account of land reforms in other areas of India. Unless otherwise noted, the description of land reform in Saurashtra is taken from this book. Additional materials, with comparable data from other regions of western India, are in G. D. Patel, The Land Problem of Reorganized Bombay State (Bombay: N. M. Tripathi Private Ltd., 1957).

TABLE 20

CARGO HANDLED (IN '000 TONS)

Year	Bedi	Sikka[a]	Navalakhi[a]	Bhavnagar	Porbandar	Veraval	Total without Okha	Okha
1936-37	143		105	304	160	175	887	
1937-38	168		109	312	221	237	1,047	
1938-39	230		101	287	212	225	1,055	
1946-47	82		77	184	185	104	632	
1947-48	180		70	212	129	149	740	
1948-49	129		168	291	140	106	834	
1949-50	196		149	300	129	119	893	
1950-51	170		136	289	158	108	861	368
1955-56	349	92	62	280	149	148	1,080	427
1960-61	346[b]	202[c]	95[d]	337[e]	142	224	1,346	583[f]

Sources: 1936-51 from Economic Survey of Saurashtra, p. 305; 1955-61 from Census of India 1961, V, Part I-A(iii), 62. Itemized composition of trade from Gujarat State Ports Traffic Review, 1960-61. Data on total imports and exports to Saurashtra ports are available in the Gazetteers, including the updating volume of 1914; in the KPA Annual Administration Reports until 1922; from Economic Survey of Saurashtra for 1931-50; in Industries of Saurashtra for 1938-48. Unfortunately it is not possible to compare these data as some are compiled in terms of tonnage, some in terms of value; some include Okha, some do not; some cover only foreign trade, some include coastal trade. Often the exact inclusiveness or exclusiveness of the data is not clearly specified.

[a]Sikka and Navalakhi are officially classed as subsidiary ports in the Bedi group of ports.
[b]Bedi includes exports of 179,000 tons of salt.   [c]Sikka includes 196,000 tons of cement exports.
[d]Navalakhi trade includes 65,000 tons of salt exports.
[e]Bhavnagar group of ports including Mahuva and other sub-ports handled 514,000 tons of trade.
[f]Okha trade includes 155,000 tons of mineral oil imports and exports of 67,000 tons of bauxite, 164,000 tons of cement, and 41,000 tons of chemicals.

TABLE 21

JOINT STOCK COMPANIES

Year	Number of Companies
Pre-1919	0
1919	1
1948	169 discovered by C. N. Vakil et al. (authors of Economic Survey of Saurashtra) plus perhaps 20 to 30 more estimated in states not investigated
1956	229 paid up capital of Rs. 13.78 crores
1961	251 paid up capital of Rs. 15.47 crores

Sources: 1948 and before, Economic Survey of Saurashtra, pp. 223-24; 1956 and 1961, Census of India 1961, V, Part I-A(iii), 48.

of the actual establishment of the United State of Saurashtra, confer[red] full occupancy rights upon the cultivators [of khalsa land] without any payment, introducing cash assessment system of land revenue and abolishing weth or forced labor."[14] This khalsa land was considered to have been state land rather than the private possession of the prince. The new state therefore succeeded to jurisdiction over it and could distribute it as seemed fit. The princes, who had relinquished control over the states, and thus the land, were granted privy purses in return, consonant with the policy of the national government.

On girasdari and barkhalidari land, the process was more complex. Girasdars held their land by virtue of historical control, usually by conquest; barkhalidars had been granted their land at some time in the past for services rendered to the ruler. Both categories paid nominal assessments to their states. These landholders argued that, unlike the princes, they actually lived in the countryside and often did at least some cultivation themselves; some parts of their land they let to tenants. Girasdari and barkhalidari land redistributed to tenants would have to be taken from people who were themselves of the countryside. Conversely, if the rural girasdars and barkhalidars were confirmed in their holdings, the tenants would have to be evicted.

---

[14]Government of Saurashtra, Memorandum Presented . . . to the Part B States . . . June 1953, p. 14.

TABLE 22

FACTORIES

Year	Number of Urban Factories	Percentage of Increase	Number of Workers	Percentage of Increase	Number of Rural Factories	Percentage of Increase	Number of Workers	Percentage of Increase
1949			623 Total Factories, Not Divided by Location					
1956	702		42,723		133		6,602	
1960	962	+37	46,031	+8	229	+72	9,885	+50

Sources: 1949, Economic Survey of Saurashtra, p. 243; 1956 and 1960, Census of India 1961, V, Part I-A(iii), 42-43.

The government met the problem by fixing maximum limits on the girasdars of three economic holdings (an economic holding being fixed at 40 acres in most parts of Saurashtra). Only girasdars who had been holding upwards of 800 acres could claim so large a plot. Most qualified for only one holding of 40 acres. Whatever land the girasdar claimed, he had to cultivate personally or with his family. Barkhalidars were limited to only one economic holding. Both categories became subject to full revenue assessment.

These laws, the Saurashtra Land Reforms Act of 1951 and the Saurashtra Barkhali Abolition Act of 1951, forced a substantial shift in the rural pattern of land control, tenure, and cultivation. The shift was most disruptive in the approximately one-third of Saurashtra's land which had been under girasdari control. Here about half of the girasdars and half of their tenants experienced a change in their land occupancy. For the most part, girasdars actually expanded direct control, evicting tenants, or at least resuming a part of the land which they had formerly leased out but which they had a right to recall under the provisions of the new laws. But the percentage of landlords in control of over 40 acres dropped from 26 per cent of the total to 17 per cent. Table 23, based on a sample of 64 villages, indicates the extent of change in the land holding pattern; the Appendix and Graph 1 reflect the land distribution.

In cases where tenants gained control over previously girasdari land, they were required to pay compensation at the admittedly high rate of Rs. 85 per acre. The Saurashtra State Government organized the Land Mortgage Bank to arrange for the purchase money facilitating the transfers. Because barkhalidari rights ordinarily had been granted for only one generation and were not hereditary, barkhali tenants who now gained land were not required to pay.

Critics from the left found the state government's actions inadequate for they treated the girasdars generously. As a class the girasdars were not a remote rentier class; they lived and worked in the villages. Clearly the government avoided root and branch reform. The new government wished to press forward economically, politically, and socially, but it recognized the conservatism of the region. The government chose to move forward only so quickly as it could carry public opinion. Part of the argument of the need for a psychological reorientation of the people to the end of the old state system and the substitution of the larger Saurashtra State unit was cited above on page 62. In addition, the government urged "that while proceeding with the work of integration, we may not lose sight of the vital factors, viz., backwardness of the State and nervousness of the people at the loss of some rights that they feel are capable of

TABLE 23

CHANGES IN THE NUMBER OF HOLDINGS OF ALL THE CATEGORIES
OF CULTIVATORS BETWEEN 1947-48 AND 1954-55

Size of Holding (acres)	Period			
	1947-48		1954-55	
	Number	Percentage	Number	Percentage
Nil	129	2.6	98	2.0
0- 5	371	7.5	404	8.1
5- 15	994	20.0	1,250	25.2
15- 25	965	19.5	1,147	23.1
25- 40	1,207	24.4	1,215	24.5
40- 60	745	15.0	573	11.5
60- 80	291	5.9	175	3.5
80-100	123	2.5	51	1.0
100-150	112	2.2	33	.7
150-200	8	.2	6	.2
200 & above	12	.2	5	.1

Source: R. R. Mishra, Effects of Land Reforms in Saurashtra (Bombay: Vora and Co., 1961), p. 61. The data here reproduced from Mishra's table seem accurate, but in his original table, he has totalled the number of holdings for both 1947-48 and for 1954-55 and both totals are grossly inaccurate. I have no idea how his errors resulted.

earning a fortune for them."[15]

The government, further, was confronted by an armed revolt of the girasdars. Politically motivated dacoity was considered responsible for 80 to 85 deaths and thousands or even lakhs of rupees looted in a series of some twelve major crimes between 1949 and 1952.[16]

If the intermediaries of Saurashtra had had their way, things in the State would have gone much the same as in Rajasthan [where no substantive change took place]. Led by the girasdars (a group of specially favored intermediaries), the landlords of Saurashtra conducted what virtually amounted to a rebellion against the Saurashtra intermediaries' abolition laws. They caused a wide breakdown of law and order, committed many dacoities and murders, and intimidated the potential beneficiaries of the land reforms among the peasantry. Some of the girasdars perhaps dreamed of causing the downfall of the Saurashtra government . . . and restoring the rule of the

---

[15] Ibid., p. 13.

[16] Rasiklal Parikh's series of articles, "Saurashtrani Raajakiya Tavaarirh: Maari Drushtia" ("The Political History of Saurashtra: From My Viewpoint"), Janmabhoomi (Bombay), June 2, 16, 23, 30, July 7, 14, 21, 28, August 4, 1968.

LAND DISTRIBUTION CHANGES FOR ALL CULTIVATORS
BETWEEN 1947-1948 AND 1954-1955

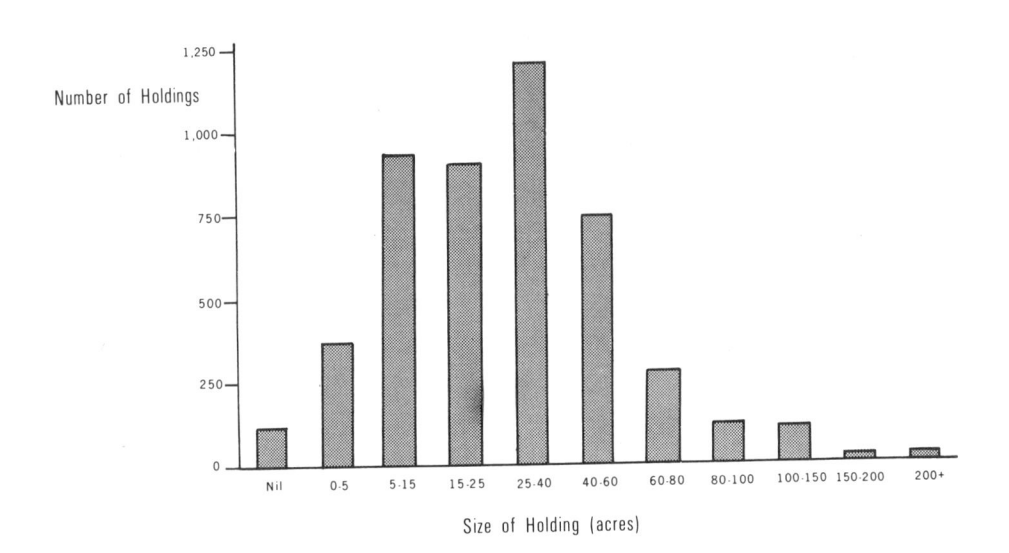

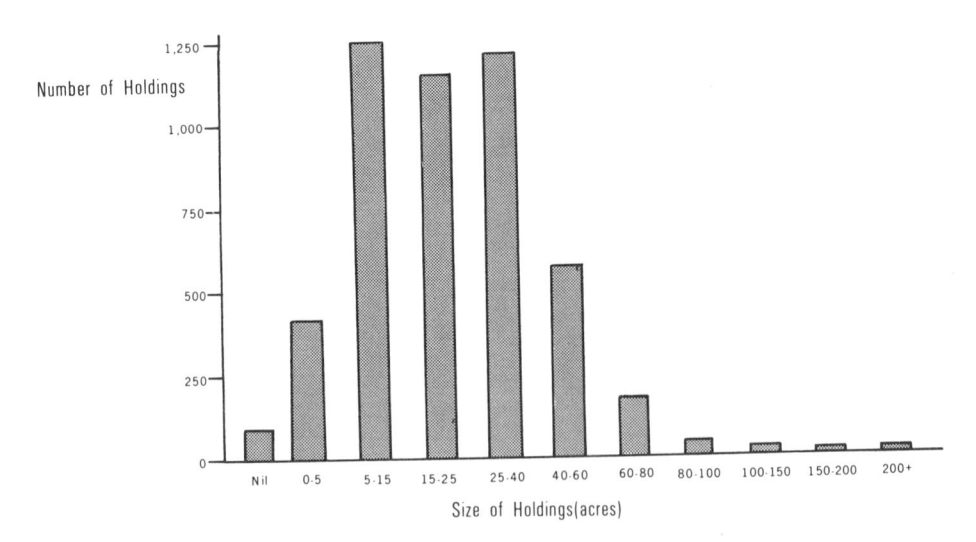

Source: R. R. Mishra, *Effects of Land Reforms in Saurashtra*, (Bombay: Vora and Co., 1961), p. 61.

Graph 1.--Distribution of Land Holdings, 1947-48 and 1954-55

princes. The girasdars, however, overestimated their strength; when the Government arrested a prince's brother--who had been associated with India's most notorious dacoit, Bhupat--and later led him through the streets in chains, the rebellion collapsed.[17]

Bhupat and the various dacoit gangs threatened the stability of the government of Saurashtra.

Girasdars, talukdars, and former rulers implicated with Bhupat included the Maharaja of Palitana, the Thakor Saheb of Dhrol, members of the ruling families of Morvi and Limbdi, and from Jamnagar, at least the Maharani. The prince's brother mentioned in Daniel Thorner's account above was the younger brother of the former ruler of Bhavnagar. He was sentenced to seven years imprisonment, later reduced to five because he turned State's witness.[18]

The Saurashtra Government, however, did not fold under pressure. It achieved both land reform and increased productivity. It compromised with girasdari demands only from a position of great, proved strength, and it did so in order to achieve a combination of justice with social harmony. By contrast, Rajasthan and Madhya Bharat, two other new states created from unions of former princely states, could not control landlord violence and political pressures; they were unable to implement effective land reforms.[19] Uttar Pradesh was similarly crippled in land reform activities by the infiltration of landlords within the Congress party.[20] The Saurashtra Government vigorously enforced land reforms and by 1953-54 it had one of the most even distributions of land holdings of any region of India (see Graph 2).

The Appendix, based on a chart appearing in the Statistical Abstract prepared by the Government of India, and Graph 2 which I prepared by applying the Lorenz Curve technique to these data, clearly indicate the comparatively greater equality of land distribution in Saurashtra as compared with other states in India. I have selected for graphing those states for which rather complete surveys had been done; these include some very large states and they include some states composed primarily of former princely states. For greater precision, I found the Gini index of inequality for Saurashtra; for its large and progressive neighbor, Bombay; and for Rajasthan, a former princely region. The index

---

[17] Daniel Thorner, The Agrarian Prospect in India (Delhi: University Press, 1956), p. 43.

[18] Parikh, "Saurashtrani Raajakiya Tavaarikh."

[19] Thorner, The Agrarian Prospect in India, pp. 29-32.

[20] Ibid., p. 50.

LAND DISTRIBUTION IN VARIOUS INDIAN STATES, 1953-54

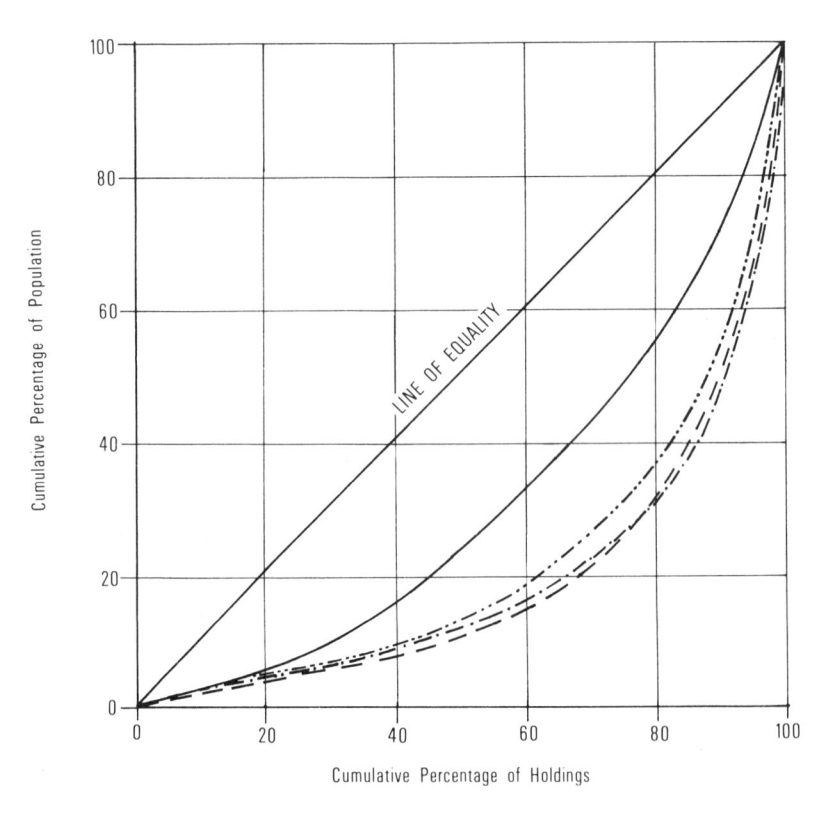

—————— Saurastra
— — —   Madras
—·—·—   Bombay
—··—··· Rajasthan

Cumulative Percentage of Population (y-axis)
Cumulative Percentage of Holdings (x-axis)

LINE OF EQUALITY

Source: Government of India, Central Statiss Statistical Organization, Statistical Abstract of India, 1956-1957 (Delhi, 1958), pp. 498-499.

Graph 2.--Distribution of Land in Various States of India, 1953-54. (Source: Government of India, Central Statistical Organization, Statistical Abstract of India, 1956-57 [Delhi, 1958], pp. 498-99).

was, respectively, .4041, .5880, and .5825.

A recent comparative study of land reforms in various parts of the world at various times in history has found that "the effects of agrarian reform, their extent and intensity, stem from the forces that create the reform in the first place more than from the reform itself."[21]

> Broad reform measures spring from dynamic social and political forces in society which push for "modernization" in general. These forces give rise to a broad scope of programs and projects--land tenure reform, education, peasant organization, enfranchisement and political participation, and interest articulation and aggregation. Agrarian reform is as much a result of development as a cause of it.[22]

Agricultural reform as part of a large package of reforms does indeed characterize the Saurashtra experience.

The Congress government recognized that the countryside needed not only a change in land control but also an increase in agricultural production. To provide this transformation it provided an entire package of economic reforms for the countryside, and began the economic restructuring of the urban-rural balance.

The package of reforms in the countryside included: increased participation by villagers in the market economy by making all agricultural revenue to the state payable in cash rather than kind; establishment of a state-run Land Mortgage Bank to assist the farmers first in buying land rights and later in financing improved farming operations; distribution of improved seeds and fertilizer;[23] and a variety of major new irrigation plans.

The tax structure of the state was revamped to shift the burden of taxation from rural to urban areas and to give the rural people control over the taxes which were collected. The change in the basis of collection effected the redress in urban-rural taxation shown in Table 24.

Even the decreased level of rural taxation still left Saurashtra with substantial tax levels in the agricultural sector. Indeed, in terms of land revenue

---

[21] R. Laporte, Jr., J. F. Petras, and J. C. Rinehart, "Agrarian Reform and Its Role in Development," Comparative Studies in Society and History, XIII, No. 4 (October, 1971), 485 (italics in original).

[22] Ibid.

[23] R. R. Mishra, Effects of Land Reforms in Saurashtra (Bombay: Vora and Co., 1961), p. 51 reflects the following distribution of improved seeds and fertilizers, 1955-56:
    12,298 maunds of cotton seeds, enough for 80,000 acres
    4,113 maunds of wheat seeds
    18,215 maunds of ammonium sulphate
    1,956 maunds of super phosphates.

TABLE 24

INCIDENCE OF TAXATION PER CAPITA

Period	Urban	Rural
1944-47	Rs. 3.8	Rs. 18.7
1953-54	17.6	10.8

Source: Government of Saurashtra, Memorandum Presented by the Government of Saurashtra to the Part B States (Special Assistance) Enquiry Committee June 1953, p. 29.

per acre, Saurashtra's rates were virtually the highest in India, exceeded only by tiny and remote Manipur and the Andaman and Nicobar Islands. In terms of agricultural revenue per capita, Saurashtra's was far and away the highest in India; it was, for example, three times that of Bombay.

The harshness of the earlier rule left the new rulers in an enviable position. They could substantially reduce revenue collection in the rural areas and still bring substantial revenues for development to the state's coffers. The wisdom of this policy of retaining substantial, but reduced, rural taxation, seems borne out by the contrary difficulties which many states in India have been experiencing in trying adequately to tax the rural sector for developmental funds. Also, it coincides with the view that a high rate of agricultural taxation levied in cash on a flat charge per acre is a great stimulus for farmers to concentrate on commercial agriculture.[24]

The introduction of panchayati raj, a form of local self-government, turned over to newly created elected bodies in the villages large portions of the taxes collected. Discretion over its expenditure passed from urban to rural hands. Table 25 indicates the growth in the number of panchayats and their disposable wealth in the decade 1950-51 to 1960-61.

The results of these innovations on the productivity of agriculture were spectacular. Acreage utilized increased steadily throughout the years of the separate Saurashtra State from about $7\frac{1}{2}$ million acres to a point where it stabil-

---

[24] Nathan Keyfitz, "Political-Economic Aspects of Urbanization in South and Southeast Asia," in The Study of Urbanization, ed. Hauser and Schnore, pp. 265-309. Cf. R. R. Mishra's explanation: "This shift towards cash crops has been helped by the fact that the cultivators are no longer on a crop-share rent basis. They are required to pay assessment in cash and the whole surplus remains with them." Effects of Land Reforms, p. 46.

TABLE 25

PANCHAYATS, 1950-51 AND 1960-61:
NUMBER, INCOME, AND EXPENDITURE

Year	Number of Panchayats	Income of Panchayats[a]	Expenditure of Panchayats[a]
1950-51	593	Rs. 892,032	Rs. 375,699
1960-61	4,329	11,511,237	8,393,044

Sources: District Census Handbooks for the six districts of the Saurashtra peninsula.

[a] Income and expenditure of panchayats omit those of Junagadh District for both 1950-51 and 1960-61. These data were not available.

ized at about $9\frac{1}{2}$ million acres, a gain of about 23 per cent. (Cf. an all-India increase in cropped area of 15 per cent between 1951 and 1961.) And production shifted sharply into commercial crops; this latter shift began under the Saurashtra State government and continued in full force after 1956. By 1961, almost two-thirds of Saurashtra's cropped area was under commercial crops; 40 per cent was under groundnut alone, which replaced cotton as the leading commercial crop of the region. The following tables (Tables 26, 27, 28, and 29) illustrate the transition during the days of the Saurashtra government and after, and also give a breakdown by different regions of the peninsula by comparing old state areas with the new districts of which they formed the cores.

The agricultural reforms were strongly oriented toward raising productivity. As we have seen, they attempted to tread lightly on vested interests unless the state itself was attacked. This stress on economic expansion coupled with social conservatism ran through the programs of the Saurashtra government. They supported small-scale rather than larger-scale industrialization and did little to foster the growth of cities. They concentrated their efforts on traditional institutions, traditional units of production, and traditional forms of entrepreneurial organization. In the villages, the socio-economic structure was largely preserved. As Schwartzberg found, "by and large the jajmani system of intra-village work allocation and reciprocal obligations appears to be more intact in Saurashtra than in any other study area."[25]

Village studies carried out in Saurashtra in conjunction with the Census of

---

[25] Schwartzberg, Occupational Structure and Level of Economic Development in India, p. 169.

TABLE 26

AREA UNDER VARIOUS CROPS (ACRES)--SAURASHTRA

Years	Total Cropped Area	Area under Food Crops	Percentage of Total Cropped Area under Food Crops	Area under Cotton	Area under Oilseeds
1949-50	7,478,300	4,207,200	56.3	614,500	1,168,000
1950-51	7,507,200	4,102,300	54.6	1,093,400	1,693,200
1951-52	7,530,900	4,198,200	55.7	950,800	1,742,700
1952-53	7,910,100	4,933,200	62.4	961,200	1,592,900
1953-54	8,747,300	4,949,200	56.6	1,022,900	1,667,300
1954-55	8,851,500	4,538,700	51.3	1,199,900	2,562,700
1955-56	8,933,200	5,050,100	56.5	1,387,100	2,051,200
1956-57	9,225,700	4,535,700	49.2	1,455,900	2,757,300
1957-58	9,137,900	4,030,800	44.1	1,326,700	3,194,700
1958-59	9,147,800	4,155,200	45.4	1,318,600	3,246,900
1959-60	9,141,100	3,621,000	39.6	1,447,400	3,779,400
1960-61	9,182,900	3,429,600	37.3	1,310,100	3,888,600
1961-62	9,294,200	3,429,100	36.9	1,256,800	3,774,200
1962-63		3,101,400		1,107,700	4,041,700
1963-64		3,134,800		1,103,200	3,477,800

TABLE 27

AREA UNDER VARIOUS CROPS (ACRES)--BHAVNAGAR

Quinquennium Ending	Total Cropped Area	Area under Food Crops	Percentage of Total Cropped Area under Food Crops	Area under Cotton	Area under Oilseeds
Bhavnagar State					
1938-39	793,661	487,789	61.5	197,473	107,005
1943-44	787,818	536,565	68.1	103,004	145,432
Bhavnagar District					
1949-50[a]	1,538,500	1,014,406	65.9	38,700	296,700
1954-55	1,632,500	984,800	60.3	135,400	401,200
1959-60	1,742,300	928,600	52.4	143,100	573,700
1961-62[b]	1,540,600	672,000	43.6	83,500	592,200

[a]One-year ending figures.   [b]Two-year ending figures.

TABLE 28

AREA UNDER VARIOUS CROPS (ACRES)--JUNAGADH

Quinquennium Ending	Total Cropped Area	Area under Food Crops	Percentage of Total Cropped Area under Food Crops	Area under Cotton	Area under Oilseeds
Junagadh State					
1938-39	550,041	228,132	41.5	142,923	153,210
1943-44	538,565	324,786	60.3	89,213	110,386
Junagadh District					
1949-50[a]	1,327,600	670,700	50.5	112,900	243,800
1954-55	1,392,400	835,000	60.0	...	332,700
1959-60	1,523,200	679,200	45.9	184,820	547,400
1961-62[b]	1,540,200	439,100[c]	28.5	103,300	779,400

[a]One-year ending figures.   [b]Four-year ending figures.

[c]Four-year figures ending 1963-64.

TABLE 29

AREA UNDER VARIOUS CROPS (ACRES)--NAWANAGAR

Quinquennium Ending	Total Cropped Area	Area under Food Crops	Percentage of Total Cropped Area under Food Crops	Area under Cotton	Area under Oilseeds
Nawanagar State					
1938-39	906,005	549,719	60.7	83,178	261,552
1943-44	827,839	577,669	69.8	26,180	214,658
Jamnagar District					
1949-50[a]	1,029,200	782,000	76.0	14,900	192,600
1954-55	1,196,200	825,500	69.0	21,100	354,100
1959-60	1,397,900	666,300	47.7	68,100	618,100
1961-62[b]	1,435,200[c]	489,500	34.1	35,700	714,400

[a]One-year ending figures.   [b]Two-year ending figures.

[c]Four-year ending 1963-64 figures.

India, 1961, affirm this view of the conservative social structure of the Saurashtrian village--even in the most economically progressive area, the Dhoraji-Upleta region of the former Gondal State.[26] In light of village studies by T. Scarlett Epstein in Mysore State, this is not surprising.[27] To the extent that villages grow wealthy by an increase in productivity per acre[28]--rather than through the creation of wholly new sources of wealth such as industry might bring--the existing social structure is reinforced.

The economy of Saurashtra began to depend on groundnut. The need to process groundnut into finished products provided new opportunities for small-scale industries, and even rural industries for preparing groundnut oil and groundnut-based fertilizers. The government responded by extending assistance to small-scale industries which were particularly suited to agro-industrial development.

In exploiting the new opportunities, the new state government showed pragmatic responses in favor of agricultural development, but it did not discriminate against the cities. Given the distinct anti-urban bias of the early Gandhian philosophy[29] and its adoption by many of the nationalist leaders in India,[30] one might have expected the Saurashtrians to have been far less moderate than they in fact proved.

Urban Development

The new leaders of Saurashtra were not, however, anti-urban. They were themselves urbanites, businessmen and professionals in the most urban

---

[26] Census of India, 1961, Vol. V--Part VI is a series of "Village Survey Monographs" on various villages in Gujarat State. Monograph No. 8--on Village Chichod in the Dhoraji Taluka of Rajkot District--is especially germane here.

[27] T. Scarlett Epstein, Economic Development and Cultural Change in South India (Manchester: Manchester University Press, 1962).

[28] In Epstein's case study, new irrigation schemes provided the means of increase; in Saurashtra, it was simply a shift from less valuable crops to more valuable.

[29] See Mohandas Karamchand Gandhi, Hind Swaraj or Indian Home Rule (Ahmedabad: Navajivan Press, 1938).

[30] Cf. Sachin Chaudhuri, "Centralization and Alternate Forms of Decentralization," in India's Urban Future, ed. Roy Turner (Berkeley: University of California Press, 1962), pp. 213-39 and Britton Harris, "Urbanization Policy in India," Papers and Proceedings of the Regional Science Association, V (1959), 192-93.

state in India. They wished to transform the cities from functioning as princely capitals to centers of industry and economic productivity, but at the same time to preserve the sense of identification which most urbanites, the new rulers thought, felt with their cities.

To serve as Industries Minister they recruited Manubhai Shah, a native Saurashtrian who had long ago left Saurashtra first to study industrial management in England and later to return to Delhi as an official of the Delhi Cloth Mills. Shah, who was later elevated to Industries Minister in the central government in New Delhi, was credited with inspiring and guiding a variety of programs for industrial development, particularly for small-scale industries for rural as well as urban locations. The state invested large sums of its regular budget as well as monies of the first five-year plan on roads, electricity, and irrigation. It established a Small Scale Industries Board and an Industrial Development Corporation--pioneering institutions later adopted throughout India-- to provide credit and technical assistance to potential small-scale businesses.

Shah and the other leaders recognized the particular importance of Rajkot as a transportation and administration center and saw it as the likely hub for small-scale industrial development as well. They established there the central economic institutions for the peninsula. In particular they sought to bolster Rajkot's economic importance before the merger of the Saurashtra State with the Government of Bombay in 1956 would terminate its political significance. In 1955, immediately after Shah moved to Delhi to help guide industrial policy for the nation, Rajkot was chosen by the central government as the location for the first industrial estate to be built in India. The industrial estate--a government built and financed center for providing worksheds and common minimum necessities such as power, water, repair shops, a bank, a canteen, etc., to a large number of private small-scale entrepreneurs[31]--was expedited with remarkable speed. The National Small Scale Industries Board adopted the construction of Industrial Estates as an appropriate technique for the development of small-scale industry for India in January, 1955. The Government of Saurashtra began construction of the first industrial estate at Rajkot in September, 1955. The first shed, of an eventual one hundred, was allotted to a small industrial unit in December, 1955.[32]

---

[31] For an account of industrial estate development in India generally, see P. C. Alexander, Industrial Estates in India (Bombay: Asia Publishing House, 1963).

[32] Ibid., pp. 17-18.

The industrial estate confirmed the transformation of Rajkot from a central place for administration, communication, transportation, and marketing to a dynamic center of increasing productivity and the promotion of innovation. The capital was becoming an industrial growth pole.[33] The construction of a new dam over the Aji River emphasized the development by securing for the city and its environs an adequate supply of water for growing population and industry.

The thrust of the new economic growth was toward agro-industries, a means of developing rural and urban areas in tandem. The need to process groundnut into finished products gave rise to new small industries for improving cultivation and for preparing groundnut oil and groundnut-based fertilizer. To supply the machinery needs of these new industries and to supply the irrigation requirements of the prospering farmers, a light engineering industry grew up specializing in the manufacture of diesel engines. By 1969, Rajkot City was producing 50,000 diesel engines per year, one-fifth of the all-India total. It supplied not only Saurashtra but other areas of India and an export market as well.[34] Table 30, concerning groundnut by-products exports indicates the extent to which the agro-industries centered on groundnut products (see page 87).

The production of food and food products employed more workers in more factories than any other branch of industry in Saurashtra (see Table 31--page 87). At least a modicum of small-scale industry was introduced to the countryside itself, as noted in Table 22, above.

The interaction between rural and urban areas also appeared in the increasing movement of farmers to the city. The new social as well as geographical mobility of the farming community is reflected in the enrollment of members of the dominant kunbi (patel, patidar) caste enrolled in urban high schools. Under most of the princes, the kunbis had been somewhat depressed in Saurashtra, particularly by comparison with their thriving caste-fellows in mainland Gujarat;[35] new policies of the Saurashtra government aided the newly-landed kunbis in particular (see Table 32).

---

[33] For a useful, if facile, comparison of central places with growth poles, see Harry W. Richardson, Elements of Regional Economics (Baltimore: Penguin Books, 1969), pp. 106-7.

[34] Interview with Jamubhai K. Modi, President, Rajkot Chamber of Commerce.

[35] Rajani Kothari and Rushikesh Maru, "Caste and Secularism in India," Journal of Asian Studies, XXV (November, 1965), 35-50.

TABLE 30

EXPORTS OF GROUNDNUT PRODUCTS ('000 TONS)

Year	Groundnut Seeds	Groundnut Cakes	Groundnut Oil
1938-39	162	54	2
1939-40	81	43	0.2
1940-41	3	12	0.1
1941-42	4	18	0.1
1942-43	5	12	0.1
1943-44	25	5	0.1
1944-45	46	2	0.05
1945-46	40	..	0.03
1946-47	8	..	negligible
1947-48	24	2	1
1949-50	Total groundnut product exports, 60,000 tons		
1954-55	34	59[a]	30
1960-61	140	196	22

Sources: 1938-50, Industries of Saurashtra, Appendix IX; 1954, Saurashtra State, Annual Administrative Report, appendices p. xlvii; 1960-61, Gujarat State Ports Traffic Review.

[a]Includes all oil cakes, not only groundnut oil.

TABLE 31

DISTRIBUTION OF FACTORIES BY MAJOR INDUSTRIES, 1961

Industry	Factories	Employees
1. Food and kindred products	385	14,203
2. Textiles and their products	67	13,161
3. Metals and metal products	190	6,786
4. Chemicals and allied products	32	4,661
5. Mining	42	3,751

Source: Census of India, 1961, V, Part I-A(iii), 42-43.

Within the urban area, the new light engineering industry was particularly profitable for the urban artisans, especially for the suthars or carpenter caste. The new state policies made it possible for them to find worksheds in the industrial estates and financing from newly created state institutions. They could overcome their shortage of capital for transforming their artisan skills into entrepreneurial productivity. Manubhai Shai, in particular, used the state fi-

TABLE 32

MEMBERS OF KUNBI (PATEL, PATIDAR) CASTE ENROLLED AT ALFRED HIGH SCHOOL, RAJKOT[a]

Year	Number of Kunbis Enrolled	Total Students Enrolled	Percentage of Kunbis
1881	2	164	1
1932-33	4	485	1
1960[b]	69	801	8.5

Sources: Class lists of the Alfred High School, Rajkot.

[a] I chose Alfred High School for this study because it is the oldest in Saurashtra and because its records were made accessible. I had wished to extend my analysis to the composition of the Raj Kumar College which became open to the public after 1939, but has remained a very exclusive and expensive boarding school. Its enrollment would be a more accurate barometer of the rising and falling fortunes of various groups in Saurashtra, but I was not permitted access to its recent records.

[b] For 1960, I took a random sample of about two-thirds of the class records. The total student enrollment was actually about 1,200.

nancial institutions to facilitate business loans to new people. More than 75 per cent of the owners of the companies in the Rajkot Engineering Association by 1969 were of the artisan castes.[36]

The new government policies did not, however, promote rapid urban growth. They were policies for balanced urban-rural regional development. The overall percentage of urbanization in the peninsula did not rise, perhaps because the former princely capitals generally lost their historic political functions. As Table 1 shows, the percentage of urbanization in Saurashtra has remained steady at 31 per cent from 1951 through 1971, while the all-India percentage rose slightly from 17 to 20 per cent.

Rapid growth in large urban areas after independence was confined to two cities: Rajkot and, to a lesser extent, Jamnagar. Both profited especially from the shift to groundnut production; other parts of the peninsula had long participated in the cotton cash crop, but these areas were not especially suited to cotton. The groundnut boom was their first opportunity to share in this market crop. Both, but particularly in Rajkot, were favored by new government policies to create from them growth points for the region. Jamnagar perhaps drew

---

[36] Interview with Gatubhai Doshi, President, Rajkot Engineering Association.

special attention economically because, as the closest significant city to the populous area of southern Pakistan and its port of Karachi, it became a military center for land, sea, and air commands of the Indian armed forces. Neither of these two cities grew to giant status. Rajkot, even as late as the census of 1971, registered only 300,000 people; a decade earlier it had not even reached 200,000.

Where urban growth did occur it was linked to industrialization, the increase of commercial activity, and the creation of productive job opportunities. This seems evident even from the marked shift in sex ratios of Saurashtra's cities. (Ideally migration data should be used to illustrate the new attractions of the cities, but in 1961, with Saurashtra merged into Gujarat State, the census data on migration is not presented in useful form.) Until 1951, the cities of Saurashtra had a higher ratio of females to males than did the rural areas. This pattern ran counter to the general pattern of Indian cities in which job opportunities induced men to come in search of work, leaving their families behind in the villages. In Saurashtra, on the contrary, men had evidently left their families in the city to seek jobs outside the peninsula. In part, at least, these men must have been the Muslim traders who were clearly an urban group, and who travelled out of Saurashtra in search of trade during substantial parts of the year leaving their families behind.[37] But, more generally and comprehensively, Saurashtrian cities had not generated adequate jobs to retain the trading community much less to attract rural people. In Chapter II we noted the extraordinary outflow of business and professional men to Bombay and other trading cities of the west coast.

Only after 1951 does the census report a shift in the sex ratios of urban as compared with rural areas. Only after 1951 does the percentage of males in the population become greater in the urban than in the rural areas. (The vast majority of the Muslim trading community left, removing one of the components of the unusual ratio. The impact of their loss was perhaps not so great as anticipated, since it should have been maximum on the 1951 census and this index was, in fact, not much affected.) The increase in urban job opportunities in small-scale industry and trade promoted by the new government attracted workers. A comparison of the sex ratios for the largest cities of the peninsula suggests that here in particular growth nodes were being created, yet even here the sex ratios were not radically different from the general sex ratios for the peninsula. Saurashtra's urban-rural pattern was a rather balanced one.

---

[37]Papanek, "Pakistan's New Industrialists and Businessmen."

TABLE 33

SEX RATIOS

Females/1,000 Males--Rajkot Division (Saurashtra Peninsula + Kutch)		
Year	Entire Division	Urban Areas Only
1901	965	985
1911	972	1,009
1921	977	1,008
1931	979	994
1941	979	984
1951	986	1,000
1961	961	945
1971	953	936

Females/1,000 Males					
Year	Rajkot City	Bhavnagar City	Jamnagar City	Junagadh City	Gondal City
1872	791	961	940	840	844
1881	847	865	978	814	938
1891	819	853	984	893	960
1901	883	898	1,012	919	957
1911	934	904	1,038	908	975
1921	927	929	994	925	1,009
1931	928	895	979	881	950
1941	959	891	972	861	982
1951	964	924	942	995	997
1961	927	916	914	911	932
1971	923	915	916	N.A.	N.A.

Conclusions

The urban system of Saurashtra, under the new political leadership was transformed to function in new ways to serve a new political philosophy which was economically progressive and socially cautious. The passing of political dominance from one urban group, the princes, to another, the Brahmin and Bania business and professional elite, engendered massive economic change in the whole regional balance of the peninsula. Saurashtra achieved the urban-rural complementary economic balance to which other areas of India aspired.[38]

It achieved this balance because the new elite, a different urban elite, held values quite different from those which had dominated the area for 140

---

[38] On the general desirability of balanced urban-rural development in India, cf. several of the articles in Desai, Grossack, and Sharms, eds., Regional Perspective of Industrial and Urban Growth: The Case of Kanpur.

years previous. The new elite owed no political favors to the old, and committed itself to quite different political, economic, and social policies. Boldly and successfully it carried through the reforms to which it had pledged itself over long decades of opposition to the former princely polity: unification, reforms in land control and tenure, the provision of amenities to rural areas, government-sponsored rural finance, and tax redistribution. The shift in the urban-rural economic and political balance resulted from these massive politically-fostered changes in the rural areas. The growth of cities was not encouraged generally, but where urban growth did occur, as in Rajkot and Jamnagar, it was largely fostered by specific government policies, particularly the creation of industrial estates. Small-scale industries were also favored through the creation of new institutions to provide both technical and financial assistance. Political decisions remained the key variable in the pattern of the growth of cities.[39]

The political phase on which this chapter focusses proved to be short-lived. The dominance of an urban elite of Brahmins and Banias in the political system of Saurashtra had ended by the mid-1960's. A newly rising group, the patidars of the countryside, had come at least to equality with the urban groups. During the first fifteen years of Independence they had profited economically from the political policies of the urban groups. Democracy increased the power of their numbers over the smaller urban elite. This passage of control of political and economic power to the hands of a dominant agricultural caste can be seen in other parts of India as well; in this Saurashtra is not unique.[40] While this further transformation of Saurashtra's urban-rural balance grows out of the political events that we analyze here, both chronologically and topically it goes beyond the limits of the present study.

---

[39] Because political decisions were so important in the development of cities, much competition among the cities developed for gaining government favors. Cf. Howard Spodek, "'Injustice to Saurashtra': A Case Study of Regional Tensions and Harmonies in India," Asian Survey, XII, No. 5 (May, 1972), 416-28.

[40] For a general discussion of this trend in India, see Donald B. Rosenthal, "Democratization, Elite Displacement, and Political Change in India," Comparative Politics, II (January, 1970), 169-201. Charles Rosen, Democracy and Economic Change in India (Berkeley: University of California Press, 1967), presents an in-depth view of the changing economic and political balance between urban and rural India since Independence. For a more universal explanation of the phenomenon, see Huntington, Political Order in Changing Societies, pp. 72-78 and 433-36.

CHAPTER IV

RAJKOT: THE CASE STUDY OF

A SINGLE CITY-STATE

> We hypothesize that the political power structure is the main variable in explaining the growth and decline of city life.
> --Gideon Sjoberg

Thus far we have examined the Saurashtra peninsula in terms of its regional integration: city with city and city with countryside. We have stressed politics as the crucial variable in both urban and regional development. Rajput and Muslim rulers, by their personal choices, set the patterns before the British came; the British chose to integrate the peninsula into one unit to a greater degree than previously; and the independent government completed that integration, establishing a new urban hierarchy in the process. This chapter focusses on the life of a single city, Rajkot, and examines its changing role in the peninsula. This account, too, stresses politics as the formative element in the city's history. The size of the population, the communications network, the number and type of industries, the extent of administrative penetration, the borders of the city and its region were all determined primarily by political decisions.

The rise in Rajkot's population rank among the cities of the peninsula reflects the increase in its regional functions over time. Table 34 ranks the largest cities in the peninsula in every decennial census from the first in 1872 through 1961. Prior to the British arrival, Rajkot was but one medium sized city among several. Its ruler held some forty villages sandwiches among the territories of the larger states of Jamnagar, Junagadh, and Gondal. As the British began to develop their Civil Station and Cantonment adjacent to the old walled city of Rajkot, and began to establish it as the communications hub of the region, beginning in the middle 1860's, the city rose in size and importance. Concomitantly it rose in population rank. Still, it trailed Bhavnagar which

TABLE 34

SIZE RANKING OF SAURASHTRA CITIES OVER 20,000, 1872-1961

1961*		1951		1941		1931		1921	
Rajkot	194,145	Bhavnagar	137,951	Bhavnagar	102,851	Bhavnagar	75,594	Bhavnagar	59,382
Bhavnagar	176,473	Rajkot	132,069	Jamnagar	71,588	Rajkot	59,112	Rajkot	45,845
Jamnagar (inc. Bedi)	148,572	Jamnagar	104,419	Rajkot	66,353	Jamnagar	55,056	Jamnagar	42,495
Surendranagar (Wadhwan)	75,706	Junagadh	62,730	Junagadh	58,111	Junagadh	39,890	Junagadh	33,221
Porbandar	75,081	Porbandar	58,824	Porbandar	48,493	Porbandar	33,383	Porbandar	28,699
Junagadh	74,298	Surendranagar-Wadhwan	57,635	Surendranagar-Wadhwan	40,589	Surendranagar-Wadhwan	31,613	Surendranagar-Wadhwan	28,111
Veraval	60,857	Dhoraji	43,787	Dhoraji	37,647	Dhoraji	29,302	Dhoraji	25,666
Morvi	50,192	Morvi	40,722	Morvi	37,048	Gondal	24,573		
Dhoraji	48,951	Veraval	40,378	Gondal	30,553	Jetpur	22,973		
Gondal	45,069	Gondal	37,046	Veraval	30,275	Veraval	21,114		
						Amreli	20,186		

1911		1901		1891		1881		1872	
Bhavnagar	60,694	Bhavnagar	56,442	Bhavnagar	57,653	Bhavnagar	47,792	Bhavnagar	35,871
Jamnagar	44,887	Jamnagar	53,844	Jamnagar	48,530	Jamnagar	39,668	Jamnagar	34,744
Junagadh	35,413	Junagadh	36,151	Junagadh	31,640	Junagadh	24,679	Junagadh	20,025
Rajkot	34,194	Rajkot	34,251	Rajkot & station	29,247	Rajkot & station	21,152	Wadhwan & station	18,448
Surendranagar-Wadhwan	24,876	Surendranagar-Wadhwan	27,478	Surendranagar-Wadhwan	24,604	Wadhwan & station	20,180	Rajkot & station	15,663
Porbandar	24,821	Dhoraji	24,815	Dhoraji	20,403			Dhoraji	15,562
Dhoraji	24,116	Porbandar	24,620					Mangrol	15,341

*Space precludes including names of cities over 20,000 since 1931. In 1941, there were a total of 14 cities over 20,000; in 1951, 18 and in 1961, there were 22.

served as the chief port and administrative capital of the largest and richest state of Saurashtra. Rajkot was becoming the central place of the peninsula, but the region was so little integrated that this function had only limited importance. After independence and merger, a regional urban hierarchy became a possibility and Rajkot immediately became its focus. By 1951, the first census after Independence showed Rajkot challenging to become the first city of the region; by 1961 it achieved the position. Graph 3 indicates two major phenomena: First the growth of the cities of Saurashtra generally and, second, the rising position of Rajkot through the urban hierarchy until it reached the position of leading city by 1961. Political decisions had created the city, elevated it, and finally raised it to the leading position in the region.

The Economic Development of Rajkot

Bardic accounts of the foundation of Rajkot note a small settlement in the place of the present city from 1259 to 1610.[1] In the latter year, Vibhaji Raju, a scion of Jamnagar's ruling family, gained possession of the tract and founded the Rajkot dynasty. During the next century, his family alternated its capital between Rajkot and the town of Sardhar twenty miles to the southeast, evidently choosing the town easier to defend at any given time. In 1720, the deputy faujdar of Junagadh, a Muslim, captured Rajkot and two years later fortified it with a wall. After twelve years of Muslim rule, Ranmalji retook Rajkot and resumed the practice of moving the capital between Rajkot and Sardhar.

Outside rulers, however, invested Rajkot with central importance. Its relative equidistance from the capitals of the major states of the peninsula and its own relative weakness, led the Gaikwad's troops to use the environs of the town as troop headquarters during mulukgiri expeditions. The British followed suit by establishing their military and administrative center on purchased land adjacent to Rajkot. Because of its significance as a manor state capital, Rajkot already had some diversity of population: ruling house, servants, banias, Muslims, Marathas, artisans, and the usual complement of functionaries to supply these groups. The British presence increased the population and the heterogeneity of the town.[2] The British added to the area Parsis, more banias, Nagars,

---

[1] For a full bardic account of Rajkot, see Shri Yaduvanshprakash, part II, pp. 36-55.

[2] P. S. Jethwa, "Rajkot ane Bhilkom" ("Rajkot and the Bhil Community") (unpublished Ph.D. dissertation, Department of Sociology, Gujarat University, Ahmedabad, 1969).

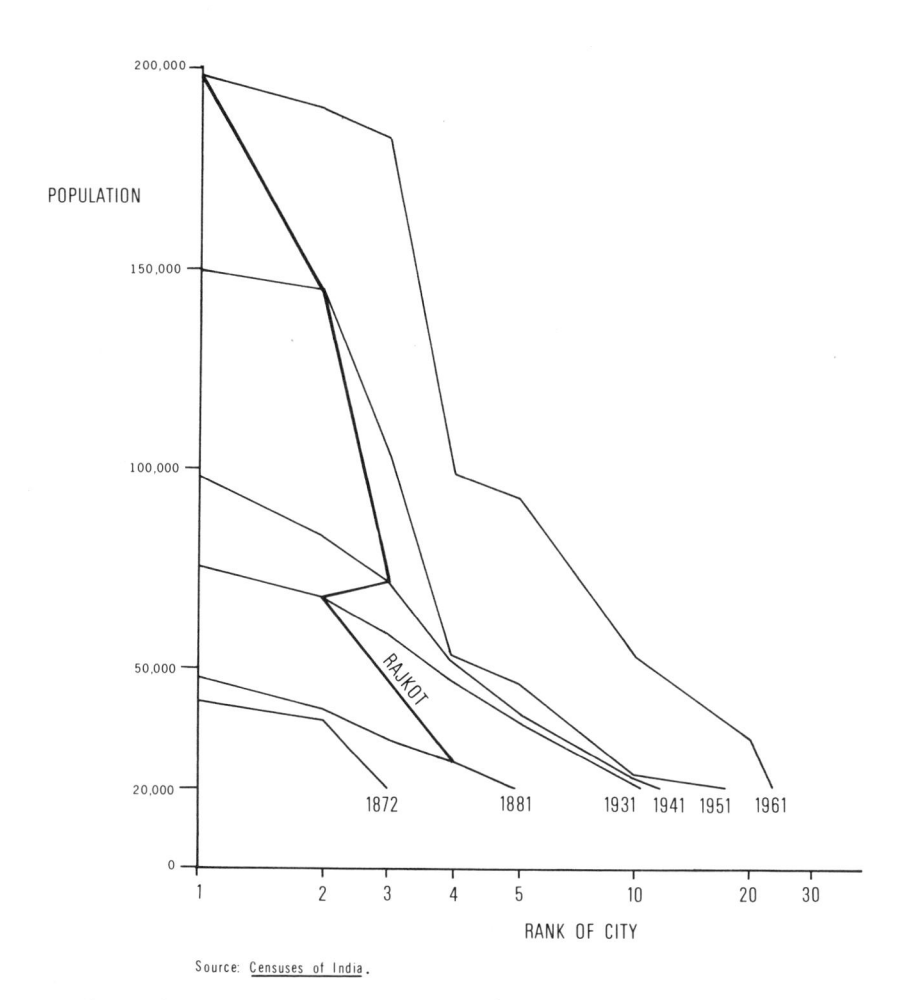

Graph 3.--Size Ranking of Saurashtra Cities, 1872-1961. (Source: Censuses of India.)

and Bhil tribals. They required troops, suppliers, railroad and road builders, and construction workers as well as administrators. After 1863, when the British began to take an active interest in affirmative administration, the population of the Civil Station multiplied, and the contiguous native town expanded <u>pari passu</u>. The British restricted the number of people in the Station allowing only those necessary for the administration and vital support services to live there. The overflow sought homes in Rajkot City. The combined British and native settlement became the fastest growing urban node in the peninsula.

Industrialization and Trade

The British-built transport, communication, and administrative network did not pay off in a rapid growth of trade and industry. Not until the 1920's and later did a general upturn appear.

TABLE 35

TRADE OF RAJKOT STATE

Year	Imports	Exports	Customs Revenue
1909-10	Rs. 2,069,962	N.A.	Rs. 48,736
1914-15	3,269,580	N.A.	57,399
1919-20	8,377,421	N.A.	100,148
1929-30	8,329,079	2,693,137	142,303

Source of data: Annual Administrative Reports of Rajkot State.

Then certain key industries began to grow (see Table 36).

What conditions changed in the 1920's to stimulate development? Apparently the most significant were changes in natural ecology and in government policies. From the turn of the twentieth century until 1920, continual outbreaks of famine and plague racked Rajkot. By 1906 the annual administrative report suggested that the state was recovering from the 1899-1901 attacks: "The time may be said to have arrived when the state can think of industries."[3] The famine of 1911 dashed that hope.[4] Seven years later, plague struck the city and forced 80 per cent of the population to flee temporarily. Not until the 1920's were ecological conditions stable enough to permit the concentration of attention

---

[3] <u>Annual Administration Report</u>, 1905-6, "Industries."

[4] DRO 7/1913-14.

TABLE 36

ORGANIZED INDUSTRIES IN RAJKOT STATE

Type	Number Functioning in Year			
	1914-15	1919-20	1929-30	1939-40
Small scale oil engine-driven flour mills	0	2	11	25
Cotton textile Mills	1	1	1	1
Spindles	0	0	7,200	10,872
Looms	112	142	500	180
B.H.P. units of electricity generated from power-house	0	0	950	1,480

Source of data: Annual Administrative Reports of Rajkot State.

and capital on industrialization.

Politically, the divisions within the peninsula inhibited Rajkot's economic development. No state wished to give advantages to another. In 1937-38, for example, Junagadh gave concessional rates on textile imports from Rajkot but, the Rajkot diwan noted, these rates would terminate as soon as Junagadh's new textile mill would be finished.[5] Competition was the normal mode. In addition, the Viramgam Customs Cordon imposed by the British also had the effect of sealing off the peninsula from exporting any finished products to the mainland. The removal of this Cordon in 1917 gave a sharp boost to trade.

Finally, Rajkot lay in the shadow of Bombay. Landlocked and backward, what attractions could Rajkot offer the investor? True it sat at the crossroads of the peninsular railroad system and served as its hub of information and administration, but Bhavnagar's port and Bombay's overall dominance were more lucrative. In 1927-28, 15-20 per cent of the state's own investments were in out-of-state companies.[6] Even the progressive Lakhaji Raj, who did much to encourage economic development in his state during his 1907-30 reign, invested heavily outside the state. In 1923-24 he held investments in the Tata Iron and

---

[5] DRO 238/1937-38, "Mill File."

[6] Annual Administration Report, 1927-28.

Steel Company; the Tata Hydro-Electric Power Supply Company, Ltd.; various Bombay companies in sugar, telephones, spinning and weaving, and agricultural equipment; in the Tapti Valley Railroad; and in the Imperial Bank.[7] If even this highly committed ruler invested heavily outside the state, what incentive could Rajkot offer to others?

Around 1920 the major obstacles to Rajkot's development began to lift. The major epidemics of plague and famine abated. The transportation and communication infrastructure began to pay off as firms from outside began to seek to establish spin-off branch plants in Rajkot. And Lakhaji Raj's government, despite its out-of-state investments, attempted to provide a climate favorable to business. Files in the District Record Office, Rajkot, include letters from Mssrs. F. C. Rustomji and Co. of Bombay as early as 1896 asking for monopoly rights in return for a spinning and weaving company. The offer, made during a minority administration, was rejected by the British-appointed administrator.[8] In 1909, Bergmann and Hoffman, Bombay and London, wrote to Rajkot and other native states as well proposing the establishment of a vegetable oil mill with machinery of Rs. 50,000.[9] A ginning factory at Sardhar was proposed as a monopoly by Bombay merchants in 1909-10.[10] In 1919-20 the Oriental Trading and Engineering Co., Calcutta, offered its services as a managing agency if Rajkot proposed initiating new industrial ventures.[11]

Rajkot actively solicited spin-offs. In 1925-26, in response to an offer by some Bombay <u>banias</u> as to the possibility of establishing a woolen mill in Rajkot, the diwan listed five major advantages for industrialization in Rajkot: 1) Local, knowledgeable exporters with outside trade contacts, 2) cheap labor, 3) a branch of the Imperial Bank to manage financial transactions, 4) important railway connections as Rajkot junction served three Kathiawad railway systems, and 5) government assistance and benefits.[12]

Behind the drive for economic development stood Lakhaji Raj, a nationalist, a friend of Gandhi, and an activist paternalistic ruler with little confidence

---

[7] DRO 7/1923-24, "Share File."

[8] DRO 1896-97 to 1898.

[9] DRO 14/1909-10.

[10] DRO 38/1909-10.

[11] DRO 12/1919-20.

[12] DRO 7/1925-26, "Mill File."

in the goodwill of the British.[14] He fostered the creation of a Rajkot Chamber of Commerce and had that organization gather data on the ports of Saurashtra and on marketing possibilities for Rajkot manufactures.[15] He attempted to contact the more prominent of the emigrants from Rajkot and have them serve as overseas agents for the state.[16] In 1920 he created a state bank.[17] In 1918-19 he built a warehouse at the Para Railway Station to help the merchants use the trains and to enable them to secure loans on their merchandise in storage.[18] He built a tramway from Rajkot to Beti and planned to extend it later.[19] In 1924, Lakhaji Raj built a small electric power house in Rajkot and began the installation of telephone equipment.[20] He supported state monopolies in flour milling and textile manufacture. After his death, merchants protested these monopolies, but, at least at the time, they seemed an appropriate mode of fostering infant industries.

The expansion of commerce and industry in Rajkot after 1947 has figured prominently in the account of the peninsula generally in Chapter III. The investments of the British era began to pay off. Moreover, as capital of the new state of Saurashtra, Rajkot's administrative functions expanded. Its population

---

[13] Cf. the argument for the importance to provincial cities of attracting spin-offs from more developed cities in Wilbur F. Thompson, "Internal and External Factors in the Development of Urban Economies," in Issues in Urban Economics, ed. Perloff and Wingo (Baltimore: Johns Hopkins Press, 1968), p. 57. Also Jane Jacobs, The Economy of Cities (New York: Vintage Books, 1970).

[14] Jayantilal Laljibhai Jobanputra, Sir Lakhajiraj Sansmarano (Rajkot: Tribhuvan Gaurishankar Vyas, 1934), a thin work, is the only biography I could locate of the ruler.

[15] DRO 1/1931-32.

[16] Cf. the letter of Rajkot State to R. J. Udani, who was already in England as the agent of the Calico Mills of Ahmedabad, urging him to serve as an agent for Rajkot products as well. DRO 1/1931-32.

[17] Annual Administration Report, 1909-10.

[18] Annual Administration Report, 1918-19.

[19] DRO 15/1915-16 for initial plans and 22/1925-26 for their continuation and rejection by Jamnagar State. DRO 1/1923-26, "Rajkot-Beti Tramway," discusses the permission of the Political Agent to Morvi State to build a branch rail line between Than Road and Chotila, ruining Rajkob's long-cherished trade prospects in Chotilla.

[20] Annual Administration Report, 1923-24.

growth and economic development paced the peninsula. In 1949 the total number of industrial workers in Rajkot City had been only 2500;[21] by 1961 it had mushroomed to 12,500.[22] The establishments in which they worked included 105 flour mills, 89 workshops making shoes and leather footwear, and 47 shops for repairing and servicing motor vehicles. The rapidly growing sector of general engineering workshops included 127 hardware manufacturing shops, 140 manufacturing and assembling machinery, and 51 manufacturing machine tools.[23] Many of the enterprises were small, with only a handful of workers, but they indicated the growth of a small industries sector. The publicly sponsored industrial estate (see Chapter III) with its 100 sheds showed the ability of the state government to enlist the support of the national government for developmental programs. Politics assisted and channeled the efforts of private business.

Rajkot and Urban-Rural Integration

The British government and the Rajkot darbar chose to isolate the countryside from the city economically and politically. They reported this policy as the wish of the peasantry. Government records report widespread fear that integration would promote urban economic exploitation of the countryside and result in political unrest. Instead the government encouraged their separation in three ways: 1) continued collection of taxes in kind rather than cash, 2) discriminatory taxation levels in the countryside, and 3) the attempt to contain political protest in the city lest it infect the countryside. Evidence of the success of the policies appears in crop patterns unresponsive to market conditions and in eyewitness reports.

The persistence of the bhagbatai system of crop share in revenue collection throughout the period limited the economic integration of the village economy with that of the capital. The British reported the system as the choice of the villagers. During the British minority administration, 1867-1876, the administrators encouraged the switch to cash collection, but only seven villages accepted it. The villagers preferred the bhagbatai system for the protection it afforded them from the moneylenders, traders, and lawyers of Rajkot City.

---

[21] Government of Saurashtra, Department of Industries and Commerce, Industries of Saurashtra (Rajkot, 1950), p. v.

[22] Census of India, Gujarat State, Vol. V, Part X-B: Special Tables on Cities and Block Directory, p. 51.

[23] Census of India 1961, Gujarat, District Census Handbook No. 2, Rajkot District, p. 162.

Cash collection might stimulate increased cash cropping, economic rationality, and business principles on the countryside, but, the farmers argued, it would also weaken village collective society. Periodically it would force them to sell when prices were low and it did not allow tax reductions in bad years. The villagers rejected the attempt of the minority administration of 1904-5 to introduce a cash settlement. In 1925-26, bhagbatai crop sharing replaced vighoti cash collections in two villages where the latter had been introduced. By this time government had long since given up its campaign to shift to cash collection and referred in the report to bhagbatai as "the natural system."[24]

Second, the weight of taxation fell heavily on the farmers while city people got off lightly. In 1928 the darbar rejected a proposal by a farm organization to limit the government share of the land produce to a maximum of 25 per cent.[25] In 1946 when the central Government of India urged a sales tax which would have fallen most heavily on the city people, the Rajkot State Government refused, for it had already experienced urban agitation against taxes and did not want to experience it again.[26] Moreover, the princely government of Rajkot never granted to the peasantry permanent occupancy rights nor rights of alienating land. Urban land, however, did carry these rights. The peasants remained until Independence tenants-at-will of the darbar.

Third, after political protest became a prominent feature in Rajkot in the late 1930's, the darbar tried to prevent its spread to the countryside. Thus during the Rajkot Satyagraha the organizers were prevented from going to the rural areas.[27]

One indication of the lack of integration of urban and rural markets was the reluctance of Rajkot State farmers to undertake cash cropping. Compare the halting response of Rajkot with the eager entrepreneurship of contiguous Gondal State in the cultivation of groundnut and cotton, the major cash crops of the state. In 1949-50, the first year for which I have data for the groundnut production for both Rajkot and Gondal talukas, approximations to the old state areas, Rajkot had only 6,158 acres of groundnut cultivation while Gondal had

---

[24]Cf. Annual Administration Reports for 1869-70 and 1872-73 in DRO 5/1873-74 and for 1873-74 in DRO 5/1874-75.

[25]DRO 5/1928-29. Although the relevant papers concern the Khedu Mahasabha, they are found in the file of the Majoor Mahamandal.

[26]DRO 94/1946.

[27]Times of India, 19 October 1938.

49,199 and Dhoraji 68,000.[28] Granted that the Gondal and Dhoraji talukas were about twice the size of the Rajkot taluka, still the ten-fold disparity in production contrasts the differential response of the two areas. More, in the first few years after independence, Rajkot caught up. Differing political policies of the Gondal and Rajkot states during the colonial period therefore seem to explain the earlier difference.

TABLE 37

ACREAGE UNDER GROUNDNUT--THREE STATE AREAS

Year	Dhoraji Taluka	Gondal Taluka	Rajkot Taluka
1949-50	68,000	49,199	6,158
1950-51	115,771	80,826	5,430
1951-52	47,499	75,728	5,340
1952-53	38,482	74,525	32,355
1953-54	22,000	67,985	28,000
1954-55	70,000	111,038	53,509

Source of data: Government of Saurashtra, Directorate of Statistics and Planning, Estimates of Area and Yield of Principal Crops (1949-50 to 1954-55) Saurashtra (Rajkot, 1955), p. 20.

Nor did Rajkot farmers specialize in the more traditional cash crop of cotton. Of 107,000 cultivated acres in Rajkot, the number under cotton cultivation is shown in Table 38.

Careful field observation by Joseph Schwartzberg in the 1950's buttresses the statistical and historical evidence of separation between urban and rural areas. After first hand evaluation of ten villages in Rajkot's vicinity, Schwartzberg concluded, "In general . . . the regional pattern is one of remarkably self-contained and self-sufficient villages, excepting for a few services obtainable only in larger towns."[29] Amplifying this account of self-sufficiency is Schwartzberg's finding "that Brahmins and the most essential artisan and serving castes

---

[28] This is particularly surprising as Rajkot was already by 1946 "Perhaps . . . the largest industrial center" in Saurashtra for crushing groundnut. DRO 95/1946-47.

[29] Joseph E. Schwartzberg, "Occupational Structure and Level of Economic Development in India: A Regional Analysis" (unpublished Ph.D. dissertation, Department of Geography, University of Wisconsin, Madison, 1960), p. 445.

TABLE 38

COTTON CULTIVATION IN RAJKOT

Year	Acres
1909-10	16,018
1910-11	17,135
1911-12	441[a]
1912-13	16,494
1913-14	22,225
1914-15	15,131
1915-16	4,798
1916-17	12,378
1917-18	32,874
1918-19	8,968
1919-20	13,959
1920-21	10,411
1921-22	4,678
1922-23	7,294
1923-24	10,606
1924-25	13,898
1925-26	16,659
1926-27	10,093
***	***
1934-35	13,548
1939-40	3,862
1944-45	63[b]

Source of data: <u>Annual Administrative Reports Rajkot State</u>.

[a] Monsoon failed; only 5" rainfall.

[b] "Grow More Food" campaign--World War II.

were found in every sample village."[30]

Occasional efforts by Lakhaji Raj's government to integrate urban and rural areas did meet with limited success. The State Bank established branches in at least half the state's villages by 1917, but little borrowing resulted (see Table 39). Postal service reached every village by 1920. By 1927-28 the government proposed telephoning the important news each day to all village communities. (The <u>Annual Administration Report</u> is not clear whether this was actually accomplished.) In 1925-26, the state hired five agricultural specialists to do both research and extension work for the benefit of the rural areas. For some time, already, the state had been selling improved seed to the villagers

---

[30] <u>Ibid.</u>, p. 441.

TABLE 39

AMOUNT BORROWED FROM VILLAGE BANKS

Year	Rupees
1916-17	34,182
1919-20	45,849
1924-25	177,769
1929-30	37,352
1934-35	26,903
1939-40	64,782
1944-45	27,047

Source of data: Annual Administrative Reports of Rajkot State.

at low prices in order to promote improved agriculture. Subsidized grain shops and public works projects established by the darbar brought streams of peasants to the city during the famines of 1878, 1899-1900, and 1911 to 1919.[31] Conversely, during the plagues of 1902-5, 1918, and 1925-26, as many as 80 per cent of Rajkot's population deserted the city at least temporarily and stayed with friends and relatives in the countryside.[32]

On the vital economic issues of markets, taxes, and crop patterns, the villages seem to have been isolated from Rajkot city. This resulted in part from peasants' choice as well as governments' decisions. The combination may have protected Rajkot's farmers from sometimes avaricious urban merchants and lawyers, but it also retarded the villagers' participation in wider market networks. The nationalist movement sought to break this isolation politically and, once in power after 1947, economically as well. Urban-rural integration, like the development of the city itself, lay largely in the hands of political leadership.

---

[31] DRO 3/1877-78, 16/1900-1901; Annual Administration Reports, 1911-12, 1912-13, 1918-19, 1925-26. In 1925-26, 174 persons died of the plague.

[32] DRO 9/1901-2 to 1903-4 and Annual Administration Report, 1902-3. In 1918-19, although the majority of the population had fled the city because of the plague, 800 people died of the flu epidemic in Rajkot. (Total population in the census of 1921 was 36,000.)

CHAPTER V

CONCLUSIONS: HISTORY AND URBAN THEORY

> Seen from a worldwide perspective, there is much to support the view that the conservative city functioning primarily as an administrative center and concerned with controlled exactions from farmers and preservation of an orderly society and hierarchy have been the much more widespread form, with the commercial city of western Europe the exception.
> --Joel Halpern, The Changing Village Community

We have stressed politics as the central determining factor in urban-rural integration in Saurashtra generally and Rajkot in particular. Now we compare this empirical case study with the major theories of the utility of cities in regional development. These theories may be grouped into four categories: growth pole, market center, communication hub, and node of specialization. The first two, in particular, stress the spatial context of economic development; they emphasize the where in planning. The theories and their elaborations often complement one another, but for analytic purposes we consider them separately.

The theory that the city may serve as a growth pole around which to organize regional economic growth has been developed most systematically by John Friedmann.[1] He has systematized a series of ideas about the city's powers of

---

[1] John Friedmann, Regional Development Policy: A Case Study of Venezuela (Cambridge: MIT Press, 1966); Urbanization, Planning and National Development (Beverly Hills: Sage Publications, 1973); and with William Alonso, eds., Regional Development and Planning: A Reader (Cambridge: MIT Press, 1964).

economic generativity and argued from them for the planting of cities in underdeveloped regions to serve as nodes for development. Friedmann cites all of the bases for cities' generativity: the city as a home of division of labor and thus of increased labor productivity; as a center for information, innovation, and diffusion of innovation; as a stimulus to economic entrepreneurship; and as a locus for political transformation. He envisions the urban node as the core of the development region. The core transmits outward to the periphery the pulse of innovation and productivity and in turn mobilizes the most productive elements of the periphery.

Friedmann calls for government action in establishing such growth poles. This separates him from those who see the city operating in a free market to accomplish regional development. It reveals the city's dependence on the political establishment. If the successful functioning of the city as growth pole depends on political resolution, then the key variable in development is in fact not the city but the political system which underlies it. The city may serve as the mechanism to promote economic development but it has little power independent of government. Ciudad Guyana, a newly developed urban growth pole which Friedmann helped plan in Venezuela's Orinoco Valley, flourished only by virtue of massive investment of resources by the government of Venezuela. The city did not create itself nor does it function in a vacuum. Plans for the use of cities as development nodes depend on governmental backing and cooperation. Without it they fail.

To the extent that Friedmann places politics as the independent variable in growth pole development, his findings are consonant with those from Saurashtra. Cities in the peninsula grew as they were chosen as capitals of rulers, and as the rulers--chieftains, British, or nationalists--chose policies of economic development. Without such backing, urban and regional economies did not develop.

While growth pole theory emphasizes the city's organizing a large region, smaller towns serve as market places for more limited areas. E. A. J. Johnson has advocated the distribution of such market towns as energizers of the countryside, as kinds of low level growth poles particularly in touch with the agricultural sector.[2] Johnson sees a well distributed network of market towns

---

[2] Johnson's general theories are best discussed in The Organization of Space in Developing Countries (Cambridge: Harvard University Press, 1970). Specifically in regard to India he wrote Market Towns and Spatial Development in India (New Delhi: National Council of Applied Economic Research, 1965).

as the means of transmitting innovation from the core to the periphery and of providing markets to tie the periphery economically with the core. Such small, local market towns provide the links which the growth pole theory requires between metropolis and peasant. Johnson bases his model on the geographer's central place theory and gives evidence of its applicability both historically and today.

Central place theory, developed in the inter-war period in Germany by Walter Christaller and August Lösch, working separately, argued that the variety of sizes of market places are spaced relatively evenly over the countryside, providing at very close intervals trade in products with low consumer thresholds--perishable foodstuffs, inexpensive articles for daily use--and at greater intervals for less frequently used goods. The regular distribution of such market places, or central places, leads to a complete urban hierarchy providing appropriate services for the hinterland of a country and linking it through the hierarchy into the entire spatial system of the country (see Figure 1). Distributed appropriately, these central places provided access to markets, sales, purchases, services, and information to everyone from the metropolitan elite to the rural farmer. Johnson applies this scheme in particular to India and argues that for the development of the Indian economy, more central place market towns must be developed at the lowest, most local level. They must be regularly spaced over the countryside to break the isolation of the rural dweller from the national market.

Johnson, however, fails to account for the political variable. He has not asked why local level market places did not spring up spontaneously. In Saurashtra, the basis seems to have been political. Local rulers established market places only where they thought them worthwhile and restricted them where they felt them threatening to the ruler's own dominance. Merchants often did not want peasants linked neatly into the market network; they wished to continue their own monopsonistic dominance. As with Friedmann, Johnson is correct to see the town--the small market town in his case--as a means of aiding economic development, but he overlooks the necessity of an underlying political commitment to begin and sustain the process.

For the market towns to serve the hinterland most effectively in promoting economic growth, political power must be re-allocated. The target population to be helped must first gain access to political power. Compare the forceful remarks by Thomas Carrol of the Inter-American Development Bank at the 1971 Rehovoth conference on urbanization in developing countries:

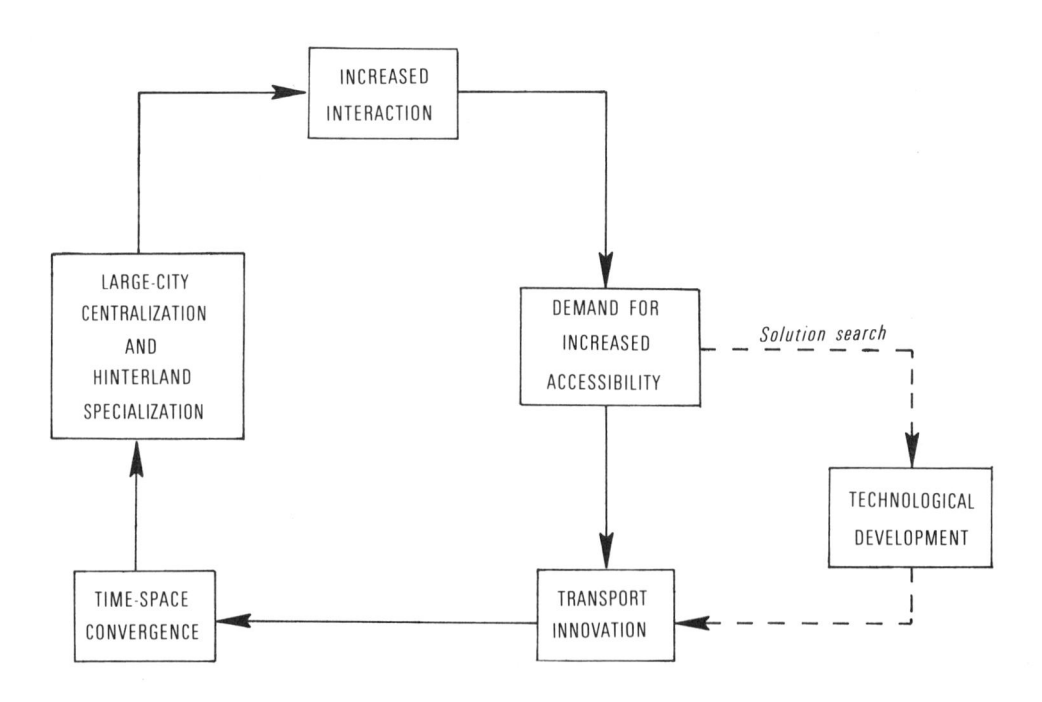

Fig. 1.--A Hierarchy of Market Centers and Market Areas. (From E. A. J. Johnson, The Organization of Space in Developing Countries, p. 20.)

> ... I am somewhat surprised that in a meeting on urbanization and development there is so little discussion of the political or power element ...
>
> Throughout the underdeveloped world there is a strand of exploitation of the cultivator. I think that we have to recognize that. The cultivator, that is the peasant who really works the land, is doubly exploited. He is exploited by the landlords, by the administrators, particularly in the feudalistic landlord-type countries. He is exploited by the market system, which is not operating in his favor, and let us be quite clear about it: He is exploited by the city, in the sense that practically all the agricultural surplus is taken away from him in some form.
>
> I feel that the attempts that have been made by governments to return some of this surplus that has been taken away through taxation, through welfare measures, through other ways now coming into being, are really a very poor substitute for strengthening the capacity of rural areas, particularly on the level of the cultivator, to retain and develop this surplus. I think we have to face the fact that its consequences are highly political, for it has to do with the power structure. Peasants are powerless. This is a proposition in most countries and if you want to speak of more balanced development you have to face this fact and deal with it. ....[3]

Our own case study supports this view. It finds that development which involved not only the city but the countryside as well was accompanied by a shift in the center of political power from aristocratic urban elites to the farmers through the mediation of the urban professional and business classes. It found that under appropriate circumstances "the city" could function effectively as market place and as growth pole, but only when the dominant political residents wished it to. The cities became most productive economically after independence as a new interest group, desiring balanced economic growth, came to power. The urban sector served the political interests which dominated it and regulated its character.

The view of the city as communications hub and therefore economically generative has also tended to ignore politics. It asserts that through communication and trade networks the city generates innovation, diffuses it, and stimulates economic growth. Major protagonists of this view include the planner Richard Meier and the geographers Torsten Hagerstrand and Allen Pred.[4] Pred's model of "The Process generating increased interaction between large

---

[3] Cited in Raanan Weitz, ed., Urbanization and the Developing Countries: Report on the Sixth Rehovoth Conference (New York: Praeger, 1973), p. 154.

[4] Richard L. Meier, "Relations of Technology to the Design of Very Large Cities," in India's Urban Future, ed. Roy Turner (Bombay: Oxford University Press, 1962), pp. 299-323; Torsten Hagerstrand, Innovation Diffusion as a Spatial Process (Chicago: University of Chicago Press, 1967); Allen Pred, The Spatial Dynamics of United States Urban-Industrial Growth, 1800-1914: Interpretive and Theoretical Essays (Cambridge: MIT Press, 1966); Pred, Urban Growth and the Circulation of Information: The United States System of Cities, 1790-1840 (Cambridge: Harvard University Press, 1973).

cities," reproduced here, also indicates how the hinterland is drawn into this process through increasing specialization and market involvement (see Figure 2). The multiplier effect which he diagrams into his model of "The circular and cumulative feedback process of urban-size growth" stimulates not only new thresholds for the city itself but also new ones for the region generally. The innovativeness and productivity of cities are expected to enrich the countryside as well (see Figure 3).

Communication theories, though noting the importance of politics, have frequently failed to consider it as a significant variable because it cannot be quantified or predicted. As Pred writes, "Despite . . . consistency, the model is not capable of accounting for all large-city population ratio adjustments and rank shifts that occurred. . . . For one thing, the model cannot provide for disturbances to stability and population ratios that were political in origin . . ."[5] He further cites James Rubin's study of three different urban communities' reactions to the building of the Erie Canal by New York City interests. Baltimore, Philadelphia, and Boston each responded quite differently as small elites in each city evaluated the challenge differently. Such political decisions affecting the use of cities in regional development are not easily amenable to systematic analysis or modelling. Indeed, studies of the building of railway networks in such diverse places as India and Mexico clearly show the political nature of the communication and transportation network and the economic class and regional biases built into it.[6] Communication hubs, like growth poles and central market places are not independent variables.

Finally, the very diverse group of thinkers who see the city as useful in generating economic growth because of its specialization and division of labor--writers as varied in time and outlook as Plato, Karl Marx, Bert Hoselitz, and Eric Lampard[7]--were certainly aware of the political dimension of urban

---

[5] Pred, Urban Growth, p. 224.

[6] Daniel Thorner, Investment in Empire: British Railway and Steam Shipping Enterprise in India, 1825-1849 (Philadelphia: University of Pennsylvania Press, 1950); Arthur Schmidt, "The Social and Economic Impact of the Railroad on Pueblas and Veracruz, Mexico, 1867-1911" (unpublished Ph.D. dissertation, Department of History, University of Indiana, Bloomington, 1973).

[7] Plato, Republic, III, 369-71; Karl Marx, Capital, trans. and ed. Frederick Engels (London: Swan, Sonnenschein, and Co., 1904), Vol. I, Part IV, chap. xiv, sec. 4; Bert Hoselitz, "Generative and Parasitic Cities," Economic Development and Cultural Change, III (1954-55), 278-96; Eric Lampard, "This History of Cities in the Economically Advanced Areas," Economic Development and Cultural Change, III (January, 1955), 81-102.

111

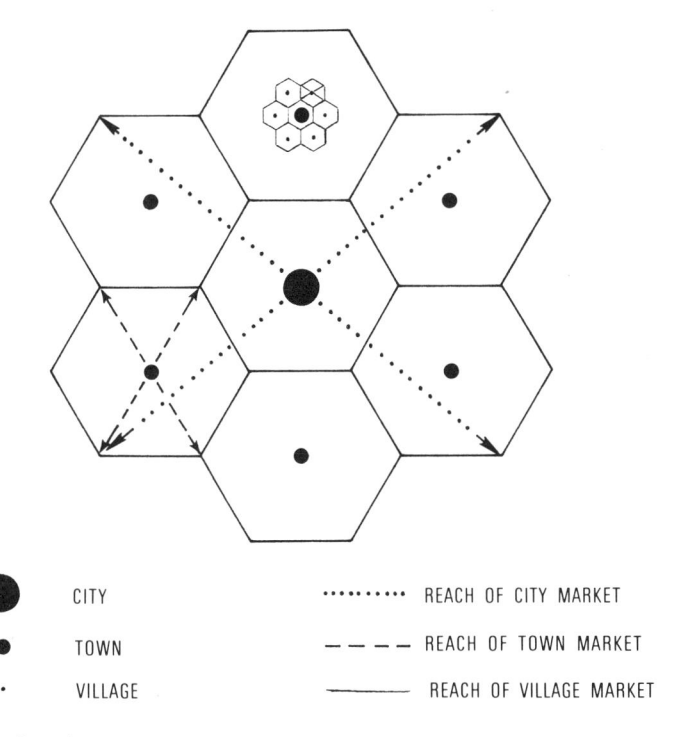

Fig. 2.--A Model of the Process Generating Increased Interaction between Large Cities. (From Allen Pred, Urban Growth and the Circulation of Information: The United States System of Cities, 1790-1840.)

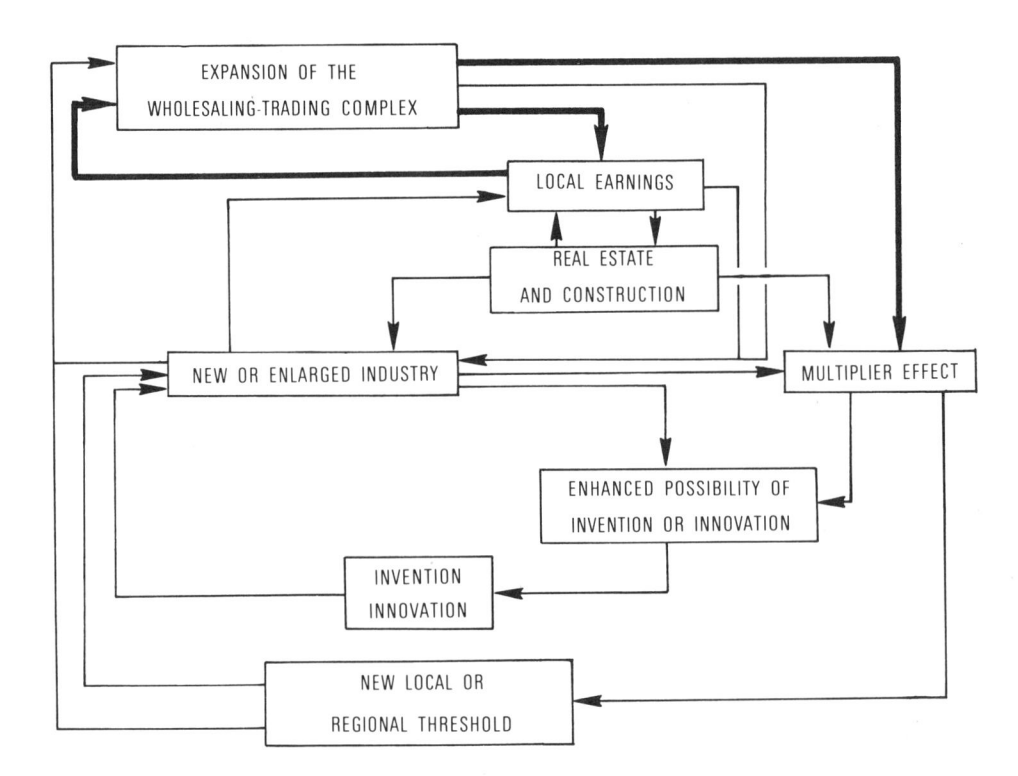

Fig. 3.--The Circular and Cumulative Feedback Process of Urban-Size Growth for the Individual American Mercantile City, 1790-1840. (From Allen Pred, Urban Growth and the Circulation of Information: The United States System of Cities, 1790-1840, p. 192.)

growth. All of them realized that the city is a tool which would produce economically according to the wishes of the dominant political interests. The city's role was determined not only by its theoretical productive potential direction.

In summary, a variety of powerful theories have suggested the use--and the creation--of cities as means of energizing regional economic development. The urban tool has been advocated by planners and implemented by governments. Most of the advocates have been aware of the political variables, but have often relegated them to a very minor place for a variety of reasons. 1) They could not be quantified or predicted. 2) The existence of a powerful nation-state capable of formulating and enforcing development policies has been taken for granted. 3) "Planners," as Uphoff and Ilchman have noted, "are unlikely, for very practical reasons, to put themselves in opposition to powerful, privileged groups within their societies."[8] James S. Coleman's remark on economists holds for planners as well: A "reason for the economist's tendency to avoid political variables is their close and more continuous link with governments as policy advisors. . . . One obvious consequence--a tendency characteristic of anyone in the establishment--is to avoid or ignore political variables."[9]

Several recent historical case studies, however, corroborate the centrality of politics in determining the economic role of cities. Paul Wheatley's study of proto-historic Shang Chinese urbanization notes progressive improvement over time in the quality of luxury goods but not in farm tools. He suggests that improved productivity was concentrated on conspicuous consumption rather than economic productivity. The recipients of urban economic productivity were, it seems, determined by the dominant political interests.[10]

Gilbert Rozman's recent diligent study of pre-modern Chinese and Japanese cities also stresses the importance of political resolutions in determining the economic role of cities.[11] Rozman groups cities into seven positions according to the functions they perform and the populations they held. They range

---

[8] Norman T. Uphoff and Warren F. Ilchman, eds., The Political Economy of Development (Berkeley: University of California Press, 1972), p. 21.

[9] James S. Coleman, "The Resurrection of Political Economy," in Political Economy of Development, ed. Uphoff and Ilchman, p. 33.

[10] Paul Wheatley, The Pivot of the Four Corners (Chicago: Aldine, 1971).

[11] Gilbert Rozman, Urban Networks in Ch'ing China and Tokugawa Japan (Princeton: Princeton University Press, 1973).

from local market places which are little more than over-grown villages, through regional centers which perform administrative as well as marketing services for a larger hinterland, up to the capital city, the apex of the urban hierarchy, which serves as the main market and administrative center for the country. The two countries' urban hierarchies developed quite differently. Japan had comparatively much greater concentration in the capital and less at the lower levels than did China because the Japanese political system was more centralized and its efficient water-based transport system enabled greater urban concentrations than in China. For Rozman, as for us, cities are not an independent variable in economic growth. They hinge on political decisions as well as on agricultural productivity and commercial skills.

Gideon Sjoberg re-emphasizes the city as dependent variable, perhaps surprisingly in view of his better known concern with technology. Sjoberg writes, "We hypothesize that the political power structure is the main variable in explaining the growth and decline of city life."[12] "The central role of cities in sustaining empires or nations helps to explain the impact of the political structure on urban growth. The political organization, to perpetuate itself, must provide a favorable climate for the development of cities. Conversely, cities cannot survive without the support of a stable, viable political system, for they are, after all, only partial systems which must import food and raw materials to sustain their population."[13] And finally Sjoberg comes close to formulating from his study very much the main conclusion of our own: "It is the primacy of political power in providing the social stability necessary for the maturation of commerce and manufacturing that is responsible for our de-emphasis of the role of purely economic or commercial factors in the rise (and diffusion and decline) of cities. These forces are significant on their own account; yet they can operate only under the aegis of a broader societal power structure."[14]

---

[12] Gideon Sjoberg, "The Rise and Fall of Cities: A Theoretical Perspective," reprinted in The City in Newly Developing Countries, ed. Gerald Breese (Englewood Cliffs, N.J.: Prentice-Hall, 1969), p. 231.

[13] Ibid., p. 222.

[14] Ibid., p. 223.

Conclusions

Two fundamentally different ways of looking at the city as a generator of regional growth emerge; they are in part related to the disciplines of the observers. Planners have tended to see the city as the engine powering the growth of regional economies. They have therfore advocated the creation of urban infrastructures in resource-rich but lagging regions and the fostering of small market towns in remote, heretofore non-market-oriented rural areas. They tend to speak of the city as an independent variable in economic development. Their plans frequently de-emphasize local political and cultural idiosyncracies.

Historians concerned with specific case studies of specific cities and regions in specific circumstances have usually shifted this emphasis.[15] They also see the city as an engine of growth, but they see the fuel for the engine in the political system. The dominant interests of the political system decide whether cities should be formed, perpetuated, and what their economic policies should be. Politics determines the distribution of wealth, profits, taxes, and future investments. Case studies, our own as well as those we have cited, of necessity pay attention to the political variable far more than planners tend to. Partly it may be a function of personality; the dynamic progress-oriented planner preferring not to be deterred in his thinking by the possibility of a political monkey wrench, an unpredictable one at that, upsetting his best laid plans; the historian, on the other hand, with the advantage of hind-sight, taking delight in being more comprehensive and presenting the contingencies in every case. The idiosyncratic nature of politics, differing from region to region, time to time, and even city to city, makes it a less likely topic for the model-oriented planner but a most appropriate one for the historian. Finally, a government-employed planner will be more circumspect in treating contemporary politics than will the privately-paid historian in writing of the past.

The two perspectives do converge, however, as planners who are also academics formulate their general models in the field. Specifically, for India, John P. Lewis, who was the director of the American aid mission to India in the mid-1960's, wrote sensitively of the difficulties of gaining political backing for schemes to use middle-sized cities to stimulate national economic growth.[16]

---

[15] Robert A. Nisbet, Social Change and History: Aspects of the Western Theory of Development (New York: Oxford University Press, 1969).

[16] John P. Lewis, Quiet Crisis in India (Bombay: Asia Publishing House, 1962).

Similarly Lloyd Rodwin's assessment of five diverse nations' attempts to use cities as levers in generating economic development spends its greatest energy explaining, and sometimes lamenting, the role of politics in supporting or sabotaging the planners' plans.[17] Most recently regional scientists engaged in planning have begun to build models of political styles into their more general planning models.[18]

In the present case study, too, there is convergence as the historian co-opts the perspectives of the planner. Saurashtra's history confirms both the validity of the planning models and the caveats of the political questions. It confirms the value of an integrated urban network and hierarchy to regional development. It confirms the significance of well distributed market towns and the importance of the city to communications and specializations. It confirms the need for a concern with the where of planning. Saurashtra's peculiar history of geographic division and political diversity makes the planner's concerns of special relevance. An understanding of the spatial aspects of development helps explain the economic changes that attended the unification of the peninsula, in limited measure under the British and more fully after Independence. It also helps explain the differential economic success of the various cities of the region. But at the same time, Saurashtra confirms the importance of political interests and decisions in determining the effectiveness of cities in promoting regional growth. A divided polity with a multitude of capital cities inhibited regional growth; a government under foreign control was not eager to see economic profit trickle down to the peasantry; but an independent government committed to unification, diffusion of opportunity, and rural as well as urban development used the cities of the peninsula quite differently and produced an impressive rate of economic growth and urban-rural integration. Cities can be generative, as the various models suggest, but this depends on the political framework which regulates their function.

---

[17] Lloyd Rodwin, Nations and Cities (Boston: Houghton Mifflin Co., 1970).

[18] Brian Berry, The Human Consequences of Urbanization (New York: St. Martin's Press, 1973).

APPENDIX

DISTRIBUTION AND SIZE OF LAND HOLDINGS--1953-54

(RELATES TO AREA OWNED, IN '000)

A. Number of Holdings

States	Less than 5 Acres	Between 5 and 10 Acres	Between 10 and 15 Acres	Between 15 and 30 Acres	Between 30 and 45 Acres	Between 45 and 60 Acres	Above 60 Acres	Total
Andhra[a]	1,767	423	178	181	50	20	26	2,645
Assam	...	...	...	...	...	...	...	...
Bihar	2,446	...	...	...	...	...	...	...
Bombay[a]	2,446	961	483	568	172	65	69	4,764
Madhya Pradesh[a]	2,648	842	386	375	105	42	60	4,458
Madras[a,h]	3,348	860	324	285	70	27	44	4,958
Orissa[b,d]	...	...	43	32	6	3	2	86
Punjab[b]	...	...	121	131	38	15	19	324
Uttar Pradesh[i]	41	5	1	2	...	...	...	48
West Bengal	...	...	...	...	...	...	...	...
Hyderabad[a,h]	897	595	385	535	190	82	114	2,798
Madhya Bharat[a]	652	323	173	193	51	18	19	1,429
Mysore[b,j]	...	...	86	84	22	8	10	210
PEPSU[b]	...	...	63	64	16	6	6	155
Rajasthan[c]	84	34	16	18	6	2	3	163
SAURASHTRA[a]	34	46	46[e]	115	60	24	18	343
Travancore-Cochin[d]	2,165	80	30[e]	4[f]	3[g]	...	...	2,282
Ajmer[a]	78	18	7	6	1	...	...	110
Bhopal[a]	39	23	17	25	9	4	6	123

States	Less than 5 Acres	Between 5 and 10 Acres	Between 10 and 15 Acres	Between 15 and 30 Acres	Between 30 and 45 Acres	Between 45 and 60 Acres	Above 60 Acres	Total
Coorg[b]	...	...	...	1	1	...	...	2
Delhi[b]	...	...	...	3	2	...	...	5
Himachal Pradesh[b]	...	...	...	6	3	1	...	10
Kutch[a]	16	17	10	17	8	4	6	78
Manipur	...	...	...	...	...	...	...	...
Tripura	...	...	...	...	...	...	...	...
Vindhya Pradesh[b]	...	...	51	64	19	8	9	151

B. Area under Land Holdings

States	Less than 5 Acres	Between 5 and 10 Acres	Between 10 and 15 Acres	Between 15 and 30 Acres	Between 30 and 45 Acres	Between 45 and 60 Acres	Above 60 Acres	Total
Andhra[a]	3,270	2,976	2,168	3,737	1,804	1,005	3,074	18,034
Assam	...	...	...	...	...	...	...	...
Bihar	...	...	...	...	...	...	...	...
Bombay[a]	5,086	6,923	6,001	11,899	6,258	3,327	7,710	47,204
Madhya Pradesh[a]	5,076	5,988	4,592	7,965	3,805	2,159	7,617	37,202
Madras[a,h]	6,592	6,006	3,952	5,853	2,553	1,399	6,194	32,549
Orissa[b,d]	...	...	521	663	210	143	170	1,707
Punjab[b]	...	...	1,515	2,743	1,378	757	2,301	8,694
Uttar Pradesh[i]	58	32	15	17	5	2	5	134
West Bengal	...	...	...	...	...	...	...	...
Hyderabad[a,h]	2,125	4,382	4,729	11,287	6,809	4,245	13,528	47,105
Madhya Bharat[a]	1,414	2,325	2,124	4,004	1,831	921	2,024	14,643
Mysore[b,j]	...	...	1,039	1,736	772	412	1,085	5,044
PEPSU[b]	...	...	788	1,321	587	288	616	3,600
Rajasthan[c]	172	244	200	382	198	107	260	1,563
SAURASHTRA[a]	100	355	579	2,528[f]	2,182[g]	1,228	1,533	8,505
Travancore-Cochin[d]	1,897	541	432[e]	126[f]	327[g]	...	...	3,323
Ajmer[a]	132	128	87	120	39	15	31	552
Bhopal[a]	62	174	212	522	321	194	770	2,255
Coorg[b]	...	...	9	23	12	9	74	127

Delhi[b]	...	...	33	37	11	3	6	90
Himachal Pradesh[b]	...	49	67	65	14	5	17	168
Kutch[a]	128	129	379	299	212	628	1,824
Manipur	...	...	...	...	...	...	...	...
Tripura	...	...	...	...	...	...	...	...
Vindhya Pradesh[b]	...	...	631	1,343	690	383	968	4,015

Source: Government of India, Central Statistical Organization, Statistical Abstract, India 1956-57 (Delhi, 1958).

[a] States where complete enumeration of holdings of all size groups was conducted.

[b] States where the census was confined to holdings of 10 acres and above.

[c] The enumeration of holdings was conducted on a sample basis of 22 selected Tehsils only.

[d] The enumeration of holdings was based on random sample.

[e] Relates to group "between 10 and 25 acres."

[f] Relates to groups "between 25 and 40 acres."

[g] Relates to group "over 40 acres."

[h] The areas have been expressed in terms of converted dry acres.

[i] The enumeration of holdings was conducted on a sample basis in 204 selected villages only.

[j] The enumeration of holdings was conducted in Government villages only.

BIBLIOGRAPHY

Archival Collections

Bombay Government Record Office, Political Department. Kathiawar. 1863.

District Record Office, Rajkot, Political Department. Rajkot State. 1867-1948.

India Office Library. Western India States Agency File.

National Archives of India. Western India States Agency File.

Newspapers

Kathiawad Times. Rajkot.

Phulchhab. Ranpur and Rajkot. Files in Phulchhab Offices, Rajkot.

Saurashtra. Ranpur. Files in Phulchhab Offices, Rajkot.

Maps

All maps are in the India Office Library, London.

Bombay Presidency, Department of the Surveyor General. "A Delineation of the Country Surveyed in the Years 1809 and 1810." Includes Bhavnagar and coast to the south.

"Map of the Western Peninsula of Guzerat commonly called Kattewar. From actual Surveys executed during the Campaigns under the Command of Lieut. Col. Walker in the Years 1807 and 1809 by Capt. Hardy."

"Map of the Province of Kattiawar. 1856."

"Kattywar Survey":
    Sheet 6: Bhavnagar and Surrounding Area (Parts of Gohelvad and Ahmedabad). 1886.
    Sheet 34: Rajkot and Environs.
    Sheet 44: Nawanagar and Surrounding Area in Halar Prant.
    "The Town and British Cantonment of Rajkot. 1873-74."

"Survey of India" covers all of Kathiawad, 1921.

Printed and Unpublished Materials

Abiodun, Josephine Olu. "Urban Hierarchy in a Developing Country." Economic Geography, XLIII, No. 4 (October, 1967), 347-67.

Adye, E. Howard. Memoir on the Economic Geology of Navanagar State. Bombay: Thacker and Co., Ltd., 1914.

Affairs of Kattywar. Part II: "General Condition and Management of Kattywar." No publication data or date; this is an exchange of views among government officials on conditions in Kathiawad around 1860. Available in District Library, Rajkot.

Aitchison, C. V., comp. A Collection of Treaties, Engagements, and Sanads Relating to India and Neighboring Countries. Vol. VI. Calcutta: Government of India, 1932.

Alexander, P. C. Industrial Estates in India. Bombay: Asia Publishing House, 1963.

Ambashankar Mahashankar. Bhavnagar Raajyano Sudhaaro. ("Reforms of the Bhavnagar State.") In Gujarati. Bhavnagar: City Central Press, 1877.

Avineri, Shlomo, ed. Karl Marx on Colonialism and Modernization. Garden City, N.Y.: Anchor Books, 1969.

Bayley, C. A. "Local Control in Indian Towns--the Case of Allahabad, 1880-1920." Modern Asian Studies, V, No. 4 (October, 1971), 289-311.

Bayley, Edward Clive. The Local Muhammedan Dynasties. Gujarat. London: W. H. Allen and Co., 1886.

Benet, Francisco. "Sociology Uncertain: The Ideology of the Rural-Urban Continuum." Comparative Studies in Society and History, VI, No. 1 (October, 1963), 1-25.

Bernier, Francois. Travels in the Mogul Empire A.D. 1656-1668. Delhi: S. Chand and Co., 1968.

Berry, Brian J. L. "City Size and Economic Development." Urbanization and National Development. Edited by Leo Jakobson and Ved Prakash. Beverly Hills: Sage Publications, 1971. Pp. 111-55.

_____. Geography of Market Centers and Retail Distribution. Englewood Cliffs, N.J.: Prentice Hall, Inc., 1969.

_____. The Human Consequences of Urbanization. New York: St. Martin's Press, 1973.

Berry, Brian J. L., et al. Indian Commodity Flows: Studies in the Spatial Structure of the Indian Economy. Chicago: University of Chicago, Department of Geography, 1966.

_____. "Policy Implications of an Urban Location Model for the Kanpur Region." Regional Perspective of Industrial and Urban Growth: The Case of Kanpur. Edited by P. B. Desai, I. M. Grossack and K. N. Sharma.

Bombay: Macmillan and Company, 1969. Pp. 203-19.

Berry, Brian J. L., and Rees, Philip H. "The Factorial Ecology of Calcutta." American Journal of Sociology, LXXIV (1969), 445-91.

Berry, Brian J. L., and Spodek, Howard. "Comparative Ecologies of Large Indian Cities." Economic Geography, XLVII, No. 1 (Supplement; June, 1971), 266-85.

Bhagvanlal Sampatraam. Saurashtra Deshno Itihaas. Vol. I. Bombay: Ganpat Krishnaajinaa Chhapaanaa, 1868.

Bhagvat Sinh Jee. Journal of a Visit to England in 1883. Bombay: Education Society's Press, 1886.

Bhatt, Tribhuvan Purushottam, comp. Sansthaan Rajkotni Directory. 4 vols. Rajkot: Rajkot State, 1929.

Bhavnagar Prajaa Parishad. Trijun Adhiveshan. Botad: N.P., 1928.

Bhavnagar State. Annual Administration Reports. 1909-10-1943-44. Bhavnagar: Bhavnagar State Printing Press, 1910-1944.

_____. Redemption of Agricultural Indebtedness in Bhavnagar State. Bombay: Times of India Press, 1934.

_____. Report on Famine Operations in the Bhavnagar State in 1899-1900. Bombay: Times of India Press, 1900.

_____. Some Important Papers Relating to the Revision Settlement of the Mahals of the Northern Division, Bhavnagar State. Bhavnagar: State Printing Press, 1936.

_____. Some Important Papers Relating to the Revision Settlement of the Mahals of the Southern Division, Bhavnagar State. 2 vols. Bhavnagar: State Printing Press, 1937.

Bhavsinhji, Sir H. H., comp. Forty Years of the Rajkumar College 1870-1910. 7 vols. London: N.P. [1911?].

The Bible.

Bloch, Marc. Feudal Society. 2 vols. Chicago: University of Chicago Press, 1964.

Bombay Presidency. Gazetteer of the Bombay Presidency. Vol. IV: Ahmedabad. Bombay, 1879.

_____. Gazetteer of the Bombay Presidency. Vol. VIII: Kathiawar. Bombay, 1884. This volume was twice updated with appendix volumes in 1907 and 1914.

Bombay Presidency, Political Department. Selections from the Records of the Bombay Government. Vols. XXXVII and XXXIX (New Series). Bombay: Government Central Press, 1856. Some volumes in this series were reprinted in 1896. In the reprint edition, the numbering of these two volumes--XXXVII and XXXIX--was reversed.

Bombay State. Master Plan for Industrialization of Bombay State. Bombay, 1960.

Bombay State, Bureau of Economics and Statistics. Statistical Abstract of Bombay State, 1957-58. Bombay, 1959.

Booth, R. B. Life and Work in India. London: J. G. Hammond, 1912.

Bose, Ashish. Urbanization in India. Bombay: Academic Books, Ltd., 1970.

_____. "The Urbanization Process in South and Southeast Asia." Urbanization and National Development. Edited by Leo Jakobson and Ved Prakash. Beverly Hills: Sage Publications, 1971. Pp. 81-109.

Bowman, J. B. Census of India, 1951: Glossary of Caste Names, Saurashtra State. Bombay: Government Central Press, 1955.

Brass, Paul. Factional Politics in an Indian State. Berkeley: University of California Press, 1966.

Breese, Gerald, ed. The City in Newly Developing Countries. Englewood Cliffs, N.J.: Prentice-Hall, Inc., 1969.

Brinton, Crane. The Anatomy of Revolution. New York: Vintage Books, 1960.

Broomfield, J. H. Elite Conflict in a Plural Society: Twentieth Century Bengal. Berkeley: University of California Press, 1968.

_____. "The Regional Elites: A Theory of Modern Indian History." The Indian Economic and Social History Review, III, No. 3 (September, 1966), 279-90.

Bulsara, Jal P. Problems of Rapid Urbanization in India. Bombay: Popular Prakashan, 1964.

Calkins, Philip B. "The Role of Murshidabad as a Regional and Subregional Center in Bengal." Urban Bengal. Edited by Richard L. Park. East Lansing: Asian Studies Center, Michigan State University, 1969. Pp. 19-28.

Center for the Study of Developing Societies. Party System and Election Studies. Bombay: Allied Publishers, 1967.

Chandra, Moti. Kaashi Kaa Itihaas. Bombay: Hindi Granth Ratnakar Private Ltd., 1962.

Chaudhuri, M. R. Indian Industries Development and Location. Calcutta: Oxford University Press, 1970.

Chaudhuri, Nirad C. The Autobiography of an Unknown Indian. Berkeley: University of California Press, 1968.

Chaudhuri, Sachin. "Centralization and the Alternate Forms of Decentralization: A Key Issue." India's Urban Future. Edited by Roy Turner. Bombay: Oxford University Press, 1962. Pp. 213-29.

Chorley, Richard J., and Haggett, Peter, eds. Models in Geography. London: Methuen, 1967.

Chudgar, P. L. Indian Princes under British Protection. London: Williams and Norgate, Ltd., 1929.

Clark, Kenneth B. Dark Ghetto. New York: Harper and Row, 1965.

Clark, Terry. Community Structure and Decision Making. San Francisco: Chandler Publishing Co., 1968.

Clune, J. H. An Historical Sketch of the Princes of India. Edinburgh: Smith, Elder, and Co., 1833.

Cohn, Bernard S. "Notes on the History of the Study of Indian Society and Culture." Structure and Change in Indian Society. Edited by Milton Singer and Bernard S. Cohn. Chicago: Aldine Publishing Co., 1968. Pp. 3-25.

──────. "Political Systems in Eighteenth Century India: The Banaras Region." Journal of the American Oriental Society, LXXXII, No. 3 (1962), 312-20.

──────. "Regions Subjective and Objective: Their Relation to the Study of Modern Indian History and Society." Regions and Regionalism in South Asian Studies: An Exploratory Study. Edited by Robert I. Crane. Durham, N.C.: Duke University Program in Comparative Studies on Southern Asia, 1967. Pp. 5-37.

──────. "Urbanization and Social Mobility in 'Early Modern' India: An Exploration." Paper prepared for the Conference of International Comparisons of Social Mobility in Past Societies, Institute for Advanced Study, Princeton, N.J., June 15-17, 1972.

Cohn, Bernard S., and Marriott, McKim. "Networks and Centers in the Integration of Indian Civilization." Journal of Social Research, I (1958), 1-9.

Commissariat, M. S. A History of Gujarat. 2 vols. Bombay: Orient Longmans, 1938 and 1957.

──────. Studies in the History of Gujarat. Bombay: Longmans, Green, 1935.

Conlon, Frank F. "The District Town as an Arena of Change in India: 1840-90." (Mimeographed.)

Coulborn, Rushton, ed. Feudalism in History. Princeton, N.J.: Princeton University Press, 1956.

Crane, Robert I., ed. Regions and Regionalism in South Asian Studies: An Exploratory Study. Durham, N.C.: Duke University Program in Comparative Studies on Southern Asia, 1967.

Cutch, Kathiawar, and Gujarat Garasia Association. Memorandum from the Cutch-Kathiawar-Gujarat Garasia Association to the Agrarian Reforms Commission of Saurashtra State. Rajkot: N.P. [1949?].

Das Gupta, Ashin. "The Merchants of Surat, c. 1700-1750." Elites in South Asia. Edited by Edmund Leach and S. N. Mukherjee. Cambridge: University Press, 1970. Pp. 201-22.

──────. "Some Attitudes among 18th Century Merchants." Ideas in History. Edited by Bisheshwar Prasad. Bombay: Asia Publishing House, 1968. Pp. 165-72.

Dave, Harikrishna Lalshankar. A Short History of Gondal. Bombay: Education Society's Press, 1889.

De Burgh, W. G. The Legacy of the Ancient World. Baltimore: Penguin Books, 1961.

De la Valette, John. An Atlas of the Progress in Nawanagar State. London: Mssrs. East and West, Ltd. [1931?].

Desai, I. P. Some Aspects of Family in Mahuva. Bombay: Asia Publishing House, 1964.

Desai, P. B.; Grossack, I. M.; and Sharma, K. N., eds. Regional Perspective of Industrial and Urban Growth: The Case of Kanpur. Bombay: Macmillan Company, 1969.

Desai, Shambhuprasad Harprasad. Junagadh and Girnar. Junagadh: Sorath Research Society, 1972.

──────. Saurashtrano Itihaas. Junagadh: Sorath Shikshan Ane Sanskruti Sangh, 1968.

Deshpande, C. D. "Visnagar: A Sample Study in the Form and Functions of Walled Towns of North Gujarat." Geographical Outlook (Ranchi), I, No. 2 (July, 1956), 1-9.

Deshpande, S. R. Report on an Enquiry into Family Budgets of Industrial Workers in Ahmedabad. N. P.: Government of India [1946?].

Deutsch, Karl W. "The Growth of Nations: Some Recurrent Patterns of Political and Social Integration." World Politics, V, No. 2 (January, 1953), 168-95.

──────. Nationalism and Social Communication: An Inquiry into the Foundations of Nationalism: Cambridge: Technology Press of the Massachusetts Institute of Technology, and New York: John Wiley and Sons, 1953.

Devanasan, Chandran D. S. The Making of the Mahatma. Delhi: Orient Longmans, 1969.

Dimock, Edward C., Jr., and Inden, Ronald B. "The City in Pre-British Bengal According to the mangala-Kavyas." Urban Bengal. Edited by Richard L. Park. East Lansing: Asian Studies Center, Michigan State University, 1969. Pp. 3-18.

The Directory of Local Self Government in Gujarat State. Bombay: All-India Institute of Local Self-Government, 1963.

Dobbin, Christine. "Competing Elites in Bombay City Politics in Mid-Nineteenth Century (1852-1883)." Elites in South Asia. Edited by Edmund Leach and S. N. Mukherjee. Cambridge: University Press, 1970. Pp. 79-94.

Dreikus, Rudolf. Equality, the Challenge of Our Times. N.P.: Private printing, 1961.

Drekmeier, Charles. Kingship and Community in Early India. Stanford: Stanford University Press, 1962.

Dumont, Louis. "The Conception of Kingship in Ancient India." Contributions to Indian Sociology, VI (December, 1962), 48-77.

Dumosia, Naoroji M. Jamnagar: A Sketch of Its Ruler and Its Administration. Bombay: Times Press, 1927.

Duncan, Otis Dudley, et al. Metropolis and Region. Baltimore: Johns Hopkins University Press, 1960.

Dwij. "Marhum Sir Prabhashankar Pattani." Bhavnagar Samachar (Weekly), XXVII (n.d.), 18-32.

Eberhard, Wolfram. Settlement and Social Change in Asia. Hong Kong: Hong Kong University Press, 1967.

Edwardes, S. J., and Fraser, L. G. Ruling Princes of India, Junagadh. Bombay: Times of India Press, 1907.

Elazar, Daniel. Cities of the Prairie. New York: Basic Books, 1970.

Epstein, A. L. "Urbanization and Social Change in Africa." Current Anthropology, VIII, No. 4 (1967), 275-96.

Epstein, T. Scarlett. Economic Development and Social Change in South India. Manchester: Manchester University Press, 1962.

Fei, Hsiao-Tung. China's Gentry: Essays in Rural-Urban Relations. Chicago: University of Chicago Press, 1953.

Fitze, Sir Kenneth. Twilight of the Maharajas. London: John Murray, 1956.

Forbes, Alexander Kinloch. Ras Mala. London: Richardson and Co., 1878.

Fox, Richard G. From Zamindar to Ballot Box: Community Change in a North Indian Market Town. Ithaca, N.Y.: Cornell University Press, 1969.

_____. Kin, Clan, Raja, and Rule: State-Hinterland Relations in Pre-Industrial India. Berkeley: University of California Press, 1971.

_____. "Rajput 'Clans' and Rurban Settlements in Northern India." Urban India: Society, Space, and Image. Edited by Richard G. Fox. Durham, N.C.: Duke University Program in Comparative Studies on Southern Asia, 1970. Pp. 167-85.

Fox, Richard G., ed. Urban India: Society, Space, and Image. Durham, N.C.:

Duke University Program in Comparative Studies on Southern Asia, 1970.

Freeman, Kathleen. Greek City-States. New York: W. W. Norton and Co., Inc., 1950.

Friedmann, John. Regional Development Policy: A Case Study of Venezuela. Cambridge: M.I.T. Press, 1966.

_____. Urbanization, Planning and National Development. Beverly Hills: Sage Publications, 1973.

Gadgil, D. R. Poona: A Socioeconomic Survey. 2 vols. Poona: Gokhale Institute of Politics and Economics, 1945 and 1952.

Galbraith, John Kenneth. The New Industrial State. New York: New American Library, 1967.

Gandhi, Mohandas Karamchand. An Autobiography. Boston: Beacon Press, 1957.

_____. The Collected Works of Mahatma Gandhi. Vol. XIII. Delhi: Publication Division, Ministry of Information and Broadcasting, Government of India, 1964.

_____. Hind Swaraj or Indian Home Rule. Ahmedabad: Navajivan Publishing House, 1938.

_____. The Indian States' Problem. Ahmedabad: Navajivan Press, 1941.

_____. Satyagraha in South Africa. Triplicane, Madras: S. Ganesan, 1928.

Garasdari Problem. Rajkot: K. S. Chandrasinhji P. Jhala [1950?].

Garner, B. "Models of Urban Geography and Settlement Location." Models in Geography. Edited by Richard J. Chorley and Peter Haggett. London: Methuen, 1967. Pp. 303-60.

Gense, J. H., and Banaji, D. R., eds. The Gaikwads of Baroda: English Documents. 10 vols. Bombay: D. B. Taraporevala and Sons [1937-1944?].

Geertz, Clifford. Islam Observed: Religious Development in Morocco and Indonesia. Chicago: University of Chicago Press, 1971.

_____. Agricultural Involution: The Process of Ecological Change in Indonesia. Berkeley: University of California Press, 1963.

Glaab, Charles N. "Historical Perspective on Urban Development Schemes." Urban Research and Policy Planning. Edited by Leo F. Schnore and Henry Fagin. Beverly Hills: Sage Publications, Inc., 1967. Pp. 197-219.

Gondal's Cherished Treasures. Gondal: Shree Bhagvat Sinhjee Golden Jubilee Committee, 1934.

Gordon, Stewart. "'Robbers' Reconsidered: Maratha Conquest and Administration of Malwa 1730-1760." Paper delivered at the Maharashtra Studies Group, Philadelphia, Pa., May, 1972.

Gottmann, Jean. Megalopolis. Cambridge: M.I.T. Press, 1961.

Govinden, P. C., comp. The Kathiawar Directory. 3 vols. Rajkot: N.P., 1921-23.

Gujarat State. Gujarat State Gazetteers. Rajkot District. Ahmedabad: Director, Government Printing, Stationery, and Publications, 1965.

_____. Gujarat State Gazetteers. Bhavnagar District. Ahmedabad: Director, Government Printing, Stationery, and Publications, 1969.

_____. Statistics of Area, Production, and Yield per Acre of Principal Crops in Gujarat State for the Period 1949-50 to 1963-64. Ahmedabad: Director, Government Printing, Stationery, and Publications, 1965.

Gujarat State, Public Works Department. Gujarat State Ports Traffic Review. 1960-61 to 1967-68. Ahmedabad: Public Works Department.

Gupta, Narajani. "Military Security and Urban Development: A Case Study of Delhi, 1857-1912." Modern Asian Studies, V, No. 1 (1971), 61-77.

Habib, Irfan. The Agrarian System of Mughal India, 1556-1707. Bombay: Asia Publishing House, 1963.

Hägerstrand, Torsten. Innovation Diffusion as a Spatial Process. Chicago: University of Chicago Press, 1967.

Halpern, Joel M. The Changing Village Community. Englewood Cliffs, N.J.: Prentice-Hall, Inc., 1967.

Hamilton, F. E. Ian. "Models of Industrial Location." Models in Geography. Edited by Richard J. Chorley and Peter Haggett. London: Methuen, 1967.

Harris, Britton. "Urbanization Policy in India." Papers and Proceedings of the Regional Science Association, V (1959), 181-207.

Harrison, Selig. India: The Most Dangerous Decades. Madras: Oxford University Press, 1960.

Hauser, Philip M. "Observations on the Urban-Folk and Urban-Rural Dichotomies as Forms of Western Ethnocentrism." The Study of Urbanization. Edited by Philip M. Hauser and Leo Schnore. New York: John Wiley and Sons, 1966. Pp. 503-17.

Hauser, Philip M., and Schnore, Leo, eds. The Study of Urbanization. New York: John Wiley and Sons, 1966.

Havell, E. B. The Ancient and Medieval Architecture of India: A Study of Indo-Aryan Civilisation. London: John Murray, 1915.

Hess, Andrew C. "The Evolution of the Ottoman Seaborne Empire in the Age of Oceanic Discoveries, 1453-1525." American Historical Review, LXXV, No. 7 (December, 1970), 1892-1919.

Hobsbawm, E. J. Primitive Rebels. New York: W. W. Norton and Co., Inc., 1965.

Holland, J. Tables of Routes and Stages through the Territories under the Presidency of Bombay Chiefly Compiled from Documents in the Office of the Quarter Master General of the Bombay Army. Bombay: Education Society's Press, 1851.

Hopkins, Edward Washburn. India Old and New. New York: Charles Scribners and Sons, 1902.

Hoselitz, Bert. "Generative and Parasitic Cities." Economic Development and Cultural Change, III (1954-55), 278-96.

―――――. "The Role of Cities in the Economic Growth of Underdeveloped Countries." Journal of Political Economy, LXI (1953), 195-208.

―――――. "Urbanization and Economic Growth in Asia." Economic Development and Cultural Change, VI (1957-58), 42-54.

Hove, Dr. Selections from the Records of the Bombay Government. Vol. XVI (New Series): Tours for Scientific and Economic Research Made in Guzerat, Kattiawar, and the Conkuns in 1787-88. Bombay: Government Central Press, 1855.

Huntington, Samuel P. Political Order in Changing Societies. New Haven, Conn.: Yale University Press, 1968.

Hurd, John, II. "Railways and the Expansion of Markets in India, 1861-1921." Paper read at Association for Asian Studies Annual Meeting, Washington, D.C., March 31, 1971.

―――――. "Some Economic Characteristics of the Princely States of India, 1901-1931." Unpublished Ph.D. dissertation, University of Pennsylvania, 1969.

Ibn Khaldun. The Muqaddimah: An Introduction to History. Translated by Franz Rosenthal. 3 vols. New York: Pantheon Books, 1958.

Government of India. Memoranda on the Native States of India 1905. Simla: Government Central Printing Office, 1905.

―――――. Memoranda on the Indian States 1921. Calcutta: Superintendent of Government Printing, 1922.

―――――. Report of the States Reorganization Commission. Delhi: Manager of Publications, 1955.

―――――. The Ruling Princes, Chiefs, and Leading Personages in the Western India States Agency. Delhi: Government of India Press, 1935.

Government of India, Central Statistical Organization. Statistical Abstract, India 1956-57. Delhi: N.P., 1958.

Government of India, Home Department. Report on Newspapers Published in the Bombay Presidency. 1876-1921. Bombay: Central Government Press, 1876-1921.

Government of India, Ministry of Health and Family Planning. Report of the

Rural-Urban Relationship Committee. 3 vols. New Delhi: The Ministry, 1966.

Government of India, Ministry of States. White Paper on Indian States. Delhi: Manager of Publications, 1950.

Indian Institute of Public Opinion. Quarterly Economic Report, Vol. VIII, No. 2 (October, 1961).

Indian Society of Agricultural Economics. Seminar on Rationale of Regional Variations in Agrarian Structure of India. Bombay: The Society, 1956.

Indian States People's Conference. Memorandum of the Indian States People's Conference Presented to the Indian State Committee. London: N.P., 1928.

Information Regarding the States in the Kathiawar Agency, and Their Leading Officials, Nobles, and Personages. September, 1923.

Jacob, George Le Grand. Western India: Before and During the Mutinies. London: Henry S. King and Co., 1872.

Jacobs, Jane. The Death and Life of Great American Cities. New York: Random House, 1961.

_____. The Economy of Cities. New York: Vintage Books, 1970.

Shri Jain Yuvak Samiti. Shri Calcutta Swetambar Sthaansavaasi (Gujarati) Jainonu Vasti-Patrak. Calcutta: N.P., 1960.

Jakobson, Leo, and Prakash, Ved. "Urbanization and Regional Planning in India." Urban Affairs Quarterly, II, No. 3 (March, 1967), 36-65.

Jakobson, Leo, and Prakash, Ved, eds. Urbanization and National Development. Beverly Hills: Sage Publications, 1971.

Jamnagar District Industrial Seminar. Souvenir. Jamnagar: N.P., 1968.

Jamnagar Factory Owners' Association. Commercial Directory of Small Scale Industrial Units and the Products Manufactured at Jamnagar. Jamnagar: The Association, n.d.

Janaki, W. A. Some Aspects of the Population Patterns in the Different Functional Groups of Towns in Gujarat. Baroda: M. S. University of Baroda Geography Research Series, 1967.

Janaki, W. A., and Ajwani, M. H. "Urban Influences and the Changing Face of a Gujarat Village." Journal of the Maharaja Sayajirao University of Baroda (Science), X, No. 2 (November, 1961), 59-87.

Jasdanwalla, Zaibun Y. Marketing Efficiency in Indian Agriculture. Bombay: Allied Publishers, 1966.

Jehangir, Sorabji. Representative Men of India. London: W. H. Allen, c. 1890.

Jethwa, P. S. "Rajkot ane Bhilkom." Unpublished Ph.D. dissertation, Gujarat University, 1969.

_____. "Sane 1900 Suhiman Rajkotnun Shaherikaran." Vaak (Saurashtra University Journal--Annual), March, 1969, pp. 56-66.

Jobanputra, Jayantilal Laljibhai. Sir Lakhajiraj Sansmarano. Rajkot: Tribhuvan Gaurishankar Vyas, 1934.

Johnson, E. A. J. "The Integration of Industrial and Agrarian Development in Regional Planning." Regional Perspective of Industrial and Urban Growth: The Case of Kanpur. Edited by P. B. Desai, I. M. Grossack and K. N. Sharma. Bombay: Macmillan Co., 1969. Pp. 171-89.

_____. Market Towns and Spatial Development in India. New Delhi: National Council of Applied Economic Research, 1965.

_____. The Organization of Space in Developing Countries. Cambridge: Harvard University Press, 1970.

Junagadh from 15 August 1947 to 20 January 1949. Junagadh: Junagadh Printing Press [1949?].

Junagadh State. The Babi Rulers of Sorath with a Short Account of Their Administration. Junagadh: N.P., 1903.

_____. Report on the Administration of the Junagadh State. 1909/10-1944/45. Junagadh: Authorized State printers from 1910 to 1945.

Kadaka, Dhanjishah Hormasji, ed. Kathiawar Local Calendar and Directory. Bombay: Education Society's Press, 1871.

_____. The Kathiawar Directory (Revised Edition). 2 vols. Rajkot: Damodar Goverdhandas Thakkar, 1886.

_____. The Kathiawar Directory. Rajkot: N.P., 1896.

Kane, P. V. History of Dharmasastra. Vol. II. Poona: Bhandarkar Oriental Research Institute, 1946.

Karbharis' Meeting. A Manual of Karbharis' Meetings of Kathiawar States (1870 to 1940). Rajkot: Under orders of Karbharis' Meeting [1940?].

Kathiawad Raajakiya Parishad. Bijun Adhiveshan tatha Kaaryavaahak Samitino Ahevaal. 1925. (No further publication data.)

Kathiawad Times--81: Special Volume for the 81st Year of Publication. Rajkot, January 26, 1968.

Katz, Elihu; Levin, Martin L.; and Hamilton, Herbert. "Traditions of Research on the Diffusion of Innovation." American Sociological Review, XXVIII, No. 2 (April, 1963), 237-52.

Katz, Michael B. "Social Structure in Hamilton, Ontario." Nineteenth Century Cities. Edited by Stephen Thernstrom and Richard Sennett. New Haven, Conn.: Yale University Press, 1969. Pp. 209-44.

Keeble, D. E. "Models of Economic Development." Models in Geography. Edited by Richard J. Chorley and Peter Haggett. London: Methuen, 1967. Pp. 243-302.

Keyfitz, Nathan. "Political-Economic Aspects of Urbanization in South and Southeast Asia." The Study of Urbanization. Edited by Philip M. Hauser and Leo F. Schnore. New York: John Wiley and Sons, Inc., 1966. Pp. 265-309.

Kincaid, Charles A. The Land of "Ranji" and "Duleep." Edinburgh and London: William Blackwood and Sons, Ltd., 1931.

Kincaid, C. A. The Outlaws of Kathiawar and Other Studies. Bombay: Times Press, 1905.

Kindersley, A. F. A Handbook of the Bombay Government Records. Bombay: Government Central Press, 1921.

Kothari, Rajani, and Mary, Rushikesh. "Caste and Secularism in India." Journal of Asian Studies, XXV (November, 1965), 33-50.

Lambert, Richard D. "The Impact of Urban Society upon Village Life." India's Urban Future. Edited by Roy Turner. Berkeley: University of California Press, 1962. Pp. 117-40.

Lampard, Eric E. "Historical Aspects of Urbanization." The Study of Urbanization. Edited by Philip M. Hauser and Leo F. Schnore. New York: John Wiley and Sons, Inc., 1966. Pp. 519-54.

Landay, Susan. "The Ecology of Islamic Cities: The Case for the Ethnocity." Economic Geography, XLVII, No. 2 (Supplement, June, 1971), 303-13.

Lapidus, Ira M. Muslim Cities in the Later Middle Ages. Cambridge: Harvard University Press, 1967.

Lapidus, Ira M., ed. Middle Eastern Cities: A Symposium on Ancient, Islamic, and Contemporary Middle Eastern Urbanism. Berkeley: University of California Press, 1969.

Laporte, R., Jr.; Petras, J. F.; and Rinehart, J. C. "Agrarian Reform and Its Role in Development." Comparative Studies in Society and History, XIII, No. 4 (October, 1971), 473-89.

Laquian, Aprodicio A. "Slums and Squatters in South and Southeast Asia." Urbanization and National Development. Edited by Leo Jakobson and Ved Prakash. Beverly Hills: Sage Publications, 1971. Pp. 183-203.

Leach, Edmund, and Mukherjee, S. N., eds. Elites in South Asia. Cambridge: University Press, 1970.

Lee-Warner, William. "Kathiawar." Journal of the Royal Society of Arts, LXI (February 28, 1913), 391-405.

_____. The Native States of India. London: Macmillan and Co., 1910.

Leonard, Karen. "The Hyderabad Political System and Its Participants." Journal of Asian Studies, XXX, No. 3 (May, 1971), 569-82.

Lewis, John P. Quiet Crisis in India. Bombay: Asia Publishing House, 1962.

Lewis, Oscar. "Further Observations on the Folk-Urban Continuum and Urbanization with Special Reference to Mexico City." The Study of Urbanization. Edited by Philip M. Hauser and Leo F. Schnore. New York: John Wiley and Sons, Inc., 1966. Pp. 491-503.

Lindholm, Sara. "The Occupational Structure of Three South Indian Cities." Seminar paper, University of Chicago, 1967. (Mimeographed.)

Long, Norton E. "The Local Community as an Ecology of Games." Urban Government. Edited by Edward C. Banfield. New York: Free Press, 1961. Pp. 400-413.

Lowi, Theodore J. The End of Liberalism: Ideology, Policy, and the Crisis of Public Authority. New York: W. W. Norton and Co., 1969.

Lynch, Kevin. The Image of the City. Cambridge: The Technology Press and Harvard University Press, 1960.

Lynch, Owen M. "Rural Cities in India: Continuities and Discontinuities." India and Ceylon: Unity and Diversity. Edited by Philip Mason. London: Oxford University Press, 1967. Pp. 142-58.

Mabogunje, A. L. Yoruba Towns. Ibadan: Ibadan University Press, 1962.

Maha Gujarat Parishad. Formation of Maha Gujarat (Memorandum Submitted to the States Re-organization Commission, Government of India). Vallabh Vidyanagar, Gujarat, 1954.

Mandelbaum, Seymour J. Boss Tweed's New York. New York: John Wiley and Sons, Inc., 1965.

Mangat, J. S. A History of the Asians in East Africa c. 1886 to 1945. Oxford: Clarendon Press, 1969.

Mannoni, O. Prospero and Caliban: The Psychology of Colonization. New York: Praeger, 1956.

Marriott, McKim, ed. Village India. Chicago: University of Chicago Press, 1955.

Marris, Peter. African City Life. Kampala: Nkanga Books [1967?].

Marx, Karl. Capital. Translated and edited by Frederick Engels. London: Swan, Sonnenschein and Co., 1904.

Maru, Rushikesh. "Fall of a Traditional Congress Stronghold." Party System and Election Studies. By Center for the Study of Developing Societies. Bombay: Allied Publishers, 1967.

Mawjee, Vishram Purshotam. The Imperial Durbar Album of the Indian Princes, Chiefs, and Zamindars. Bombay: Lakshmi Art Printing Works, n.d.

McDonald, Ellen E. "City-Hinterland Relations and the Development of a Regional Elite in 19th Century Bombay."

McGee, Terry. "Catalysts or Cancers? The Role of Cities in Asian Society." Urbanization and National Development. Edited by Leo Jakobson and Ved Prakash. Beverly Hills: Sage Publications, 1971. Pp. 157-81.

Jhaverchand Meghani. "A Lamp of Humanity." Indian Literature, IX, No. 3 (July-September, 1966), 56-70 and X, No. 2 (April-June, 1967), 32-65.

_____. Sorathi Bahaarvatiyaa. 4 vols. Ahmedabad: Gurjara Grantharatna Kaaryaalaya, 1929.

_____. Sorath, Taran Vahetan Paani. Ahmedabad: Gurhara Grantharatna Kaaryaalaya, 1937.

Mehta, Damodardas Bhagvandas, comp. A Collection of Kathiawad Political Agency Circulars, Orders, Notifications, Rules, Etc., from 1864-1905. 2 vols. Ahmedabad: United Printing Press [1905?].

Mehta, Gordhandas Nagardas. Saurashtra Itihaas Darshan. Palitana: B. P. Press, 1937.

Mehta, J. M. A Study of the Rural Economy of Gujarat Containing Possibilities of Reconstruction. Baroda: Baroda State Press, 1930.

Mehta, Kowshikaram Vighrhararam, ed. Gowrishankar Udayashankar Oza. Bombay: Times of India Press, 1903.

Meier, Richard L. Developmental Features of Great Cities of Asia. Vol. II: Japanese, Chinese, and Indian. Working Paper No. 124. Berkeley: College of Environmental Design, University of California, May, 1970.

_____. "Exploring Development in Great Asian Cities: Seoul." American Institute of Planners Journal, November, 1970, pp. 378-92.

_____. "The Metropolis and the Transformation of Resources." Bulletin of the Atomic Scientists, XXVI, No. 5 (May, 1970), 2-5 and 36-37.

_____. "Policies for Planning Rural-Urban Migration: Urban Villages Reconsidered." Behavior in New Environments. Edited by Eugene B. Brody. Beverly Hills: Sage Publications, 1970. Pp. 395-404.

Mellor, John W., et al. Developing Rural India: Plan and Practice. Ithaca, N.Y.: Cornell University Press, 1968.

Menon, V. P. The Story of the Integration of the Indian States. Bombay: Orient Longmans, 1961.

Metcalf, Thomas R. The Aftermath of Revolt: India 1857-1870. Princeton, N.J.: Princeton University Press, 1964.

_____. "From Raja to Landlord: The Oudh Talukdars, 1850-1870." Land Control and Social Structure in Indian History. Edited by Robert Eric Frykenberg. Madison: University of Wisconsin Press, 1969. Pp. 123-41.

Metcalf, Thomas R. "Social Effects of British Land Policy in Oudh." Land Control and Social Structure in Indian History. Edited by Robert Eric Frykenberg. Madison: University of Wisconsin Press, 1969. Pp. 143-62.

Mirat-i-Ahmadi. A Persian History of Gujarat. Translated by M. F. Lokhandwala. Baroda: Oriental Institute, University of Baroda, 1965.

Mishra, R. R. Effects of Land Reforms in Saurashtra. Bombay: Vora and Co., 1961.

Moore, Barrington, Jr. Social Origins of Dictatorship and Democracy. Boston: Beacon Press, 1965.

Moreland, W. H. The Agrarian System of Moslem India. Delhi: Oriental Books Reprint Corp., 1968.

_____. India at the Death of Akbar. Delhi: Atma Ram and Sons, 1962.

Morris, H. S. The Indians in Uganda. Chicago: University of Chicago Press, 1968.

Morris, Morris D., et al. Indian Economy in the Nineteenth Century: A Symposium. Delhi: Delhi School of Economics, 1969.

Motiwala, B. N. Karsondas Mulji: A Biographical Study. Bombay: Karsondas Mulji Centenary Celebration Committee, 1937.

Mukherjee, S. N. "Class, Caste, and Politics in Calcutta, 1815-38." Elites in South Asia. Edited by Edmund Leach and S. N. Mukherjee. Cambridge: University Press, 1970. Pp. 32-78.

Murphey, Rhoads. "The City as a Center of Change in Western Europe and China." Annals of the Association of American Geographers, XLIV, No. 4 (December, 1954), 349-62.

_____. "Traditionalism and Colonialism: Changing Urban Roles in Asia." Journal of Asian Studies, XXIX, No. 1 (November, 1969), 67-84.

_____. "The Treaty Ports and China's Modernization: What Went Wrong." The Chinese City between Two Worlds. Edited by Mark Elvin and G. William Skinner. Stanford: Stanford University Press, forthcoming.

_____. "City and Countryside as Ideological Issues: India and China." Comparative Studies in Society and History, XIV, No. 5 (June, 1972), 250-67.

Naqvi, Hameeda Khattoon. Urban Centres and Industries in Upper India 1556-1803. Bombay: Asia Publishing House, 1968.

Nath, Kidar. The Cantonment Laws in India. Lahore: University Book Agency, 1937.

Nawanagar Chamber of Commerce. Chamber Bulletin. Vol. I, No. 1.

_____. Industrial Information Center Souvenir. Jamnagar, 1965.

_____. Memorandum to National Shipping Board. February 22, 1969.

Nawanagar Chamber of Commerce. Memoranda on Port Development.

Nawanagar State. Annual Administration Reports. 1909/10-1943/44. Nawanagar, 1910-1944.

Nawanagar State and Its Critics. Bombay: Times of India Press, 1929.

Neale, Walter. Economic Change in Rural India: Land Tenure and Reform in Uttar Pradesh, 1800-1955. New Haven, Conn.: Yale University Press, 1962.

_____. "Land Is to Rule." Land Control and Social Structures in Indian History. Edited by Robert Eric Frykenberg. Madison: University of Wisconsin Press, 1969. Pp. 3-16.

Newcombe, Vernon Z. "Gandhinagar: A New Capital for the State of Gujarat (India)." Journal of the Town Planning Institute, LIV, No. 3 (March, 1968), 123-28.

Nightingale, Pamela. Trade and Empire in Western India, 1784-1806. Cambridge: Cambridge University Press, 1970.

Nilsson, Sten. European Architecture in India, 1750-1850. London: Faber and Faber, 1968.

Nisbet, Robert A. Social Change and History: Aspects of the Western Theory of Development. New York: Oxford University Press, 1969.

Opler, Morris E. "The Extensions of an Indian Village." Journal of Asian Studies, XVI, No. 1 (November, 1956), 5-10.

Oza, Kevalram C. Reconstruction of Life and Polity in Kathiawar States. Rajkot: Published by author, 1946.

Pahl, R. E. "Sociological Models in Geography." Models in Geography. Edited by Richard J. Chorley and Peter Haggett. London: Methuen, 1967. Pp. 217-42.

Pandit, D. P. Earning One's Livelihood in Mahuva. Bombay: Asia Publishing House, 1965.

Pannikar, K. M. A Survey of Indian History. Bombay: Asia Publishing House, 1963.

Papanek, Hanna. "Pakistan's New Industrialists and Businessmen: Focus on the Menons." Paper delivered at the Conference on Occupational Cultures in Changing South Asia, University of Chicago, May 15-16, 1970.

Parikh, Narhari D. Sardar Vallabhbhai Patel. 2 vols. Ahmedabad: Navajivan Publishing House, 1953 and 1956. (Gujarati edition published 1950).

Parikh, Rasiklal. "Saurashtrani Raajakya Tavaarikh: Maari Drushtie." A series of 54 articles appearing in the newspaper Janmabhoomi, December 24, 1967-December 29, 1968.

Park, Richard L., ed. Urban Bengal. East Lansing: Asian Studies Center,

Michigan State University, 1969.

Parker, Donald Dean. <u>Local History: How to Gather It, Write It, and Publish It</u>. New York: Social Science Research Council, 1944.

Pearson, Michael Naylor. "Commerce and Compulsion: Gujarati Merchants and the Portuguese System in Western India, 1500-1600." Unpublished Ph.D. dissertation, University of Michigan, 1971.

Perloff, Harvey S., and Wingo, Lowdon, Jr., eds. <u>Issues in Urban Economics</u>. Baltimore: Johns Hopkins Press, 1968.

Phadnis, Urmila. <u>Towards the Integration of Indian States, 1919-1947</u>. Bombay: Asia Publishing House, 1968.

Philips, C. H. <u>The Evolution of India and Pakistan, 1858-1947</u>. London: Oxford University Press, 1962.

Phillimore, R. H. <u>Historical Records of the Survey of India</u>. 4 vols. Dehra Dun: Survey of India, 1945.

Pirenne, Henri. <u>Medieval Cities: Their Origins and the Revival of Trade</u>. Princeton, N.J.: Princeton University Press, 1952.

Plato. <u>The Republic</u>. Translated with an introduction and notes by Francis MacDonald Cornford. New York: Oxford University Press, 1945.

Pocock, David F. "Sociologies--Urban and Rural." <u>Contributions to Indian Sociology</u>, IV (1960), 83-91.

Prakash, Ved. "Land Policies for Urban Development." <u>Urbanization and National Development</u>. Edited by Leo Jakboson and Ved Prakash. Beverly Hills: Sage Publications, 1971. Pp. 205-24.

Prasad, Bisheshwar, ed. <u>Ideas in History</u>. Bombay: Asia Publishing House, 1968.

Pred, Allan. <u>The Spatial Dynamics of U.S. Urban-Industrial Growth, 1800-1914: Interpretive and Theoretical Essays</u>. Cambridge: M.I.T. Press, 1966.

_____. <u>Urban Growth and the Circulation of Information: The United States System of Cities, 1790-1840</u>. Cambridge: Harvard University Press, 1973.

Qanungo, Bhupen. "A Study of British Relations with the Native States of India, 1858-1962." <u>Journal of Asian Studies</u>, XXVI, No. 2 (February, 1967), 251-65.

Rahasthanik Court. <u>Hak Patrak Decisions of the Rajasthanik Court Kathiawad 1873-1897</u>. 10 vols. Rajkot: Damodardas Printing House, 1934.

Ramusack, Barbara Nell. "Indian Princes as Imperial Politicians 1914-1939." Unpublished Ph.D. dissertation, University of Michigan, 1969.

Ranchodji Amarji. <u>Tarikh-i-Sorath</u>. Bombay: Education Society's Press [1882?].

Rajkot Municipal Corporation. "Report on the Development Plan of Rajkot." Rajkot, 1965. (Typescript.)

Rajkot State. The Annual Administration Report of the Rajkot State. 1901/10-1944-45. Rajkot, 1910-1945.

Raol, Virbhadra Singhji K. "Social Organization of Rajputs in Saurashtra." Unpublished Ph.D. dissertation, University of Bombay, 1969.

Ratnamanirao Bhimrao. Gujaratnun Patnagar: Ahmedabad. Ahmedabad: Gujarat Sahitya Sabha, 1929.

――――――. Khambaatno Itihaas. Cambay: Cambay State, 1934.

Raychaudhuri, Tapan. "A Reinterpretation of Nineteenth Century Indian Economic History?" Indian Economy in the Nineteenth Century: A Symposium. Edited by Morris D. Morris, et al. Delhi: Delhi School of Economics, 1969. Pp. 77-100.

Redford, Arthur. Labor Migration in England, 1800-1850. Manchester: Manchester University Press, 1964.

Richardson, Harry W. Elements of Regional Economics. Baltimore: Penguin Books, 1969.

Ridker, Donald G. "Prospects and Problems of Agriculture in the Kanpur Region." Regional Perspective of Industrial and Urban Growth: The Case of Kanpur. Edited by P. B. Desai, I. M. Grossack, and K. N. Sharma. Bombay: Macmillan and Co., 1969. Pp. 55-72.

Rodwin, Lloyd. Nations and Cities: A Comparison of Strategies for Urban Growth. Boston: Houghton Mifflin Co., 1970.

Rörig, Fritz. The Medieval Town. Berkeley: University of California Press, 1967.

Rosen, George. Democracy and Economic Change in India. Berkeley: University of California Press, 1967.

Rosenthal, Donald B. "Deurbanization, Elite Displacement, and Political Change in India." Comparative Politics, II (January, 1970), 169-201.

――――――. The Limited Elite: Politics and Government in Two Indian Cities. Chicago: University of Chicago Press, 1970.

Roy, R. "Administration of Industrial Development in U.P." Regional Perspective of Industrial and Urban Growth: The Case of Kanpur. Edited by P. B. Desai, I. M. Grossack, and K. N. Sharma. Bombay: Macmillan and Co., 1969. Pp. 262-79.

Rozman, Gilbert. Urban Networks in Ch'ing China and Tokugawa Japan. Princeton: Princeton University Press, 1973.

Rudolph, Lloyd I., and Rudolph, Suzanne Hoeber. "The Political Role of India's Caste Association." Pacific Affairs, XXVIII, No. 3 (1955), 235-53.

Rudolph, Lloyd I., and Rudolph, Suzanne Hoeber. "Rajputana under British Paramountcy: The Failure of Indirect Rule." Journal of Modern History, XXXVIII (June, 1966), 138-60.

Saletore, B. A. Main Currents in the Ancient History of Gujarat. Baroda: The Maharaja Sayajirao University of Baroda, 1960.

Sankalai, Hasmukh D. The Archaeology of Gujarat. Bombay: Natvarlal and Co., 1941.

Sarkar, Sir Jadunath. Mughal Administration. Calcutta: M. C. Sarkar and Sons, Ltd., 1935.

Government of Saurashtra. Annual Administration Report. 1950/51-1954/55. Rajkot: Government of Saurashtra, 1951-1955.

_____. Memorandum Presented by Government of Saurashtra to the Part B States (Special Assistance) Enquiry Committee, June 1953.

_____. Memorandum on the Relationship between the Government of India and the Government of Part B States.

_____. Memorandum on the State Bank of Saurashtra Submitted by the Government of Saurashtra to the Government of India, 1955.

_____. Saurashtra Government Gazette.

_____. The Saurashtra Town Planning Act, 1955 (Act No. XII of 1955). Rajkot, 1955.

_____. Traffic Review of Saurashtra Ports, 1955-56. Bhavnagar, 1956.

Government of Saurashtra, Department of Industries. Report on Cottage and Small Scale Industries of Japan by the Saurashtra Cottage Industries Board's Delegation to Japan. Rajkot, 1951.

Government of Saurashtra, Department of Industries and Commerce. Industries of Saurashtra. Rajkot, 1950.

Government of Saurashtra, Directorate of Statistics and Planning. Draft Second Five Year Plan 1956-1961. General Review. Rajkot, 1955.

_____. Estimates of Area and Yield of Principal Crops (1949-50 to 1954-55) Saurashtra. Rajkot, 1955.

Government of Saurashtra, Director of Information and Publicity. Fifth Year of Freedom in Saurashtra. Rajkot: Saurashtra Government Press, 1952.

_____. Seventh Year of Freedom in Saurashtra. Rajkot: Saurashtra Government Press, 1954.

_____. Sixth Year of Freedom in Saurashtra. Rajkot: Saurashtra Government Press, 1953.

Schwartzberg, Joseph E. "Occupational Structure and Level of Economic Development in India: A Regional Analysis." Unpublished Ph.D. dissertation,

University of Wisconsin, Madison, Wisc., 1960. In 1961 this was published as Monograph No. 4 by the Census of India under the auspices of Government of India, Ministry of Home Affairs.

Seal, Anil. The Emergence of Indian Nationalism. Cambridge: Cambridge University Press, 1968.

Shah, A. M. "Political System in Eighteenth Century Gujarat." Enquiry, I, No. 1 (Spring, 1964), 83-95.

Shah, A. M., and Shroff, R. G. "The Vahivanca Barots of Gujarat: A Caste of Genealogists and Mythographers." Traditional India: Structure and Change. Edited by Milton Singer. Philadelphia: American Folklore Society, 1959. Pp. 40-70.

Shah, Maneklal H. Jam the Great. Nadiad: Gujarat Times Office, 1934.

Shils, Edward. The Intellectual between Tradition and Modernity: The Indian Situation. The Hague: Mouton and Co., 1961.

Singer, Milton. "Beyond Tradition and Modernity in Madras." Comparative Studies in Society and History, XIII, No. 2 (April, 1971), 160-95.

_____. When a Great Tradition Modernizes. New York: Praeger, 1972.

Singer, Milton, ed. Traditional India: Structure and Change. Philadelphia: American Folklore Society, 1959.

Singer, Milton, et al. "Urban Politics in a Plural Society: A Symposium." Journal of Asian Studies, XX, No. 3 (May, 1961), 265-97.

Singh, Kashi Nath. "The Territorial Basis of Medieval Town and Village Settlement in Eastern Uttar Pradesh, India." Annals of the Association of American Geographers, LVIII, No. 2 (June, 1968), 203-20.

Singh, Nihal. Shree Bhagvat Sinhjee: The Maker of Modern Gondal. Gondal: N.P., 1934.

Singh, R. L. Banaras: A Study in Urban Geography. Banaras: Nand Kishore and Bros., 1955.

Sjoberg, Gideon. The Preindustrial City. New York: Free Press, 1960.

_____. "The Rise and Fall of Cities: A Theoretical Perspective." International Journal of Comparative Sociology, IV, No. 2 (September, 1963), 107-20.

Smith, Adam. An Inquiry into the Nature and Causes of the Wealth of Nations. New York: Modern Library, 1937.

Smith, Page. As a City upon a Hill. New York: Alfred A. Knopf, 1966.

Soja, Edward W. The Geography of Modernization in Kenya. Syracuse Geographical Series, No. 2. Syracuse, N.Y.: Syracuse University Press, 1968.

Solanki, S. R. "Saurashtra Regional Plan." Unpublished thesis, School of Planning, London, 1954.

Sopher, David E. "Pilgrim Circulation in Gujarat." Geographical Review, LVIII, No. 3 (July, 1968), 392-425.

Sovani, N. V. "The Analysis of 'Over-Urbanization.'" Economic Development and Cultural Change, XII, No. 2 (January, 1964), 113-22.

──────. Urbanization and Urban India. Bombay: Asian Publishing House, 1966.

Spodek, Howard. "'Injustice to Saurashtra': A Case Study of Regional Tensions and Harmonies in India." Asian Survey, XII, No. 5 (May, 1972), 416-28.

──────. "The 'Manchesterization' of Ahmedabad." Economic Weekly, XVII, No. 11 (March 13, 1965), 483-90.

──────. "On the Origins of Gandhi's Political Methodology: The Heritage of Kathiawad and Gujarat." Journal of Asian Studies, XXX, No. 2 (February, 1971), 361-72.

──────. "Traditional Culture and Entrepreneurship." Economic and Political Weekly, VI, No. 8 (February 22, 1969), M-27--M-31.

Steed, Gittel. "Notes on an Approach to a Study of Personality Formation in a Hindu Village in Gujarat." Village India. Edited by McKim Marriott. Chicago: University of Chicago Press, 1955. Pp. 102-44.

Stein, Burton. "The Economic Function of a Medieval South Indian Temple." Journal of Asian Studies, XIX, No. 2 (February, 1960), 163-76.

──────. "Integration of the Agrarian System of South India." Land Control and Social Structure in Indian History. Edited by Robert Eric Frykenberg. Madison: University of Wisconsin Press, 1969. Pp. 175-216.

──────. "The Segmentary State in Indian History." Paper delivered at the American Historical Association Annual Meeting, New York City, December, 1971.

Stein, Maurice R. The Eclipse of a Community. Princeton, N.J.: Princeton University Press, 1960.

Surati, N. M. Garasdari Problem. Rajkot: K. S. Chandrasinhji P. Jhala [1950?].

Tanna, Ratilal. Alag Saurashtra Shaa Maate? Rajkot: N.P. [1966?].

Taylor, Thomas. Memoirs of the Life and Writings of the Right Reverend Reginald Heber, D.D. Late Lord Bishop of Calcutta. London: John Hatchard and Sons, 1836.

Techno-Economic Survey of Gujarat. New Delhi: National Council of Applied Economic Research, 1963.

Terssac, Mademoiselle Urbanie de. A Travers le Kattiawar. (Reprint in India Office Library--no further publication data.)

Thakkar, Narandas P. Jamnagar Kaal ane Aaj. Jamnagar: N.P., 1966.

Thernstrom, Stephen, and Sennett, Richard, eds. Nineteenth Century Cities. New Haven, Conn.: Yale University Press, 1969.

Thompson, Wilbur R. "Internal and External Factors in the Development of Urban Economics." Issues in Urban Economics. Edited by Harvey S. Perloff and Lowdon Wingo, Jr. Baltimore: Johns Hopkins University Press, 1968.

──────. A Preface to Urban Economics. Baltimore: Johns Hopkins University Press, 1965.

Thorner, Daniel. The Agrarian Prospect in India. Delhi: University Press, 1956.

──────. "Feudalism in India." Feudalism in History. Edited by Rushton Coulborn. Princeton, N.J.: Princeton University Press, 1956.

Timms, Duncan. The Urban Mosaic: Towards a Theory of Residential Differentiation. Cambridge: University Press, 1971.

Tod, James. Annals and Antiquities of Rajasthan. 2 vols. London: George Routledge and Sons, Ltd., 1914.

──────. Travels in Western India. London: Wm. H. Allen and Company, 1839.

Trivedi, A. B. Kathiawar Economics. Bombay: N.P., 1943.

Trivedi, Ramanlal K. Bhavnagar State Census, 1931, Part I: "Report." Bhavnagar, 1932.

Turner, Roy, ed. India's Urban Future. Berkeley: University of California Press, 1962.

Urban-Rural Differences in Southern Asia: Some Aspects and Methods of Analysis. Delhi: UNESCO Research Center on Social and Economic Development in Southern Asia, 1964.

Social Research on Small Industries in India. Delhi: UNESCO Research Center on Social and Economic Development in Southern Asia, 1963.

U.S. Government. House Document No. 294. Second Annual Report of the Appalachian Region Commission, 1967. Washington, D.C.: Government Printing Office, 1968.

Upadhyaya, J. R. Mahatma Gandhi as a Student. Delhi: Publications Division, Ministry of Information and Broadcasting, Government of India, 1965.

──────. Mahatma Gandhi: A Teacher's Discovery. Vallabh Vidyanagar: Sardar Patel University, 1969.

Uphoff, Norman T., and Illchman, Warren F., eds. The Political Economy of Development. Berkeley: University of California Press, 1972.

Vakil, C. N.; Lakdawala, D. T.; and Desai, M. B. Economic Survey of Saurashtra. Bombay: School of Economics and Sociology, University of Bombay, 1953.

Ved, Vallabhdas S., comp. Government Resolutions in Giras and Political Cases. 5 vols. Ahmedabad: Dwarkadas V. Ved, 1910.

Vidich, Arthur J., and Bensman, Joseph. Small Town in Mass Society. Princeton, N.J.: Princeton University Press, 1968.

Wade, Richard. The Urban Frontier. Chicago: University of Chicago Press, 1959.

Wakefield, Sir Edward. Past Imperative: My Life in India 1927-1947. London: Chatto and Windus, 1966.

Waley, Daniel. The Italian City-Republics. New York: McGraw-Hill Book Co., 1969.

Warner, Sam Bass, Jr. The Private City: Philadelphia in Three Periods of Its Growth. Philadelphia: University of Pennsylvania Press, 1968.

Warner, W. Lloyd, et al. Yankee City. 1 vol. abridged ed. New Haven, Conn.: Yale University Press, 1963.

Watson, J. W. Statistical Account of Bhavnagar. Bombay: Education Society's Press [1884?].

_____. Statistical Account of Dhrangadhra. Bombay: Education Society's Press, 1884.

_____. Statistical Account of Junagadh. Bombay: Bombay Gazette Steam Press, 1884.

_____. Statistical Account of Nawanagar. Bombay: Education Society's Press, 1879.

_____. Statistical Account of Porbandar. Bombay: Education Society's Press, 1879.

Weber, Max. The City. New York: Free Press, 1958.

_____. The Religion of India. Glencoe, Ill.: Free Press of Glencoe, 1958.

Weiner, Myron. Party Building in a New Nation. Chicago: University of Chicago Press, 1967.

_____. The Politics of Scarcity. Chicago: University of Chicago Press, 1962.

Weitz, Raanan, ed. Urbanization and the Developing Countries: Report on the Sixth Rehovoth Conference. New York: Praeger, 1973.

Wellisz, Stanislaw. "Economic Development and Urbanization." <u>Urbanization and National Development</u>. Edited by Leo Jakobson and Ved Prakash. Beverly Hills: Sage Publications, 1971. Pp. 39-55.

Western India States Agency. <u>List of Princes, Chiefs, and Talukdars Exercising Jurisdiction in the Western India States Agency with the Names of Their Next of Kin</u>. Rajkot: N.P., 1929.

_____. <u>The Ruling Princes, Chiefs, and Leading Personages in the Western India States Agency</u>. Rajkot: N.P., 1928.

Wheeler, Sir Mortimer. <u>The Civilizations of the Indus Valley and Beyond</u>. New York: McGraw-Hill Book Co., 1966.

Wilberforce-Bell, H. <u>The History of Kathiawad</u>. London: William Heinemann, 1916.

Wild, Roland. <u>The Biography of His Highness Shri Sir Ranjitsinhji</u>. London: Rich and Cowan, Ltd., 1934.

Williams, L. F. Rushbrook. <u>The Black Hills: Kutch in History and Legend: A Study in Indian Local Loyalties</u>. London: Weidenfeld and Nicolson, 1958.

Williams, William Appleman. <u>The Great Evasion</u>. Chicago: Quadrangle Books, 1964.

Wolf, Eric R. <u>Peasants</u>. Englewood-Cliffs, N.J.: Prentice-Hall, Inc., 1966.

Wood, John R. "The Political Integration of British and Princely Gujarat: The Historical-Political Variable in Indian State Politics." Unpublished Ph.D. dissertation, Columbia University, 1971.

<u>Shree Yaduvansh Prakaash</u>. I have no further exact publication details on this bardic history of the Jadeja Rajput States. The place of publication is presumably Jamnagar or Rajkot. The approximate year of publication is 1934.

Yajnik, Jhaverilal Umiashankar. <u>Gaorishankar Udayashankar, C.S.I.</u> Bombay: Education Society's Press [1890?].

Zeigler, Norman. "The Rajput State of Mewar: Adaptation to a Scarce Resource Base." Paper delivered at University of Chicago Indian History Seminar, April 14, 1970.

THE UNIVERSITY OF CHICAGO
DEPARTMENT OF GEOGRAPHY
RESEARCH PAPERS (Lithographed, 6×9 Inches)

(Available from Department of Geography, The University of Chicago, 5828 S. University Ave., Chicago, Illinois 60637. Price: $6.00 each; by series subscription, $5.00 each.)

106. SAARINEN, THOMAS F. *Perception of the Drought Hazard on the Great Plains* 1966. 183 pp.
107. SOLZMAN, DAVID M. *Waterway Industrial Sites: A Chicago Case Study* 1967. 138 pp.
108. KASPERSON, ROGER E. *The Dodecanese: Diversity and Unity in Island Politics* 1967. 184 pp.
109. LOWENTHAL, DAVID, et al. *Environmental Perception and Behavior.* 1967. 88 pp.
110. REED, WALLACE E. *Areal Interaction in India: Commodity Flows of the Bengal-Bihar Industrial Area* 1967. 210 pp.
112. BOURNE, LARRY S. *Private Redevelopment of the Central City: Spatial Processes of Structural Change in the City of Toronto* 1967. 199 pp.
113. BRUSH, JOHN E., and GAUTHIER, HOWARD L., JR. *Service Centers and Consumer Trips: Studies on the Philadelphia Metropolitan Fringe* 1968. 182 pp.
114. CLARKSON, JAMES D. *The Cultural Ecology of a Chinese Village: Cameron Highlands, Malaysia* 1968. 174 pp.
115. BURTON, IAN; KATES, ROBERT W.; and SNEAD, RODMAN E. *The Human Ecology of Coastal Flood Hazard in Megalopolis* 1968. 196 pp.
117. WONG, SHUE TUCK. *Perception of Choice and Factors Affecting Industrial Water Supply Decisions in Northeastern Illinois* 1968. 96 pp.
118. JOHNSON, DOUGLAS L. *The Nature of Nomadism* 1969. 200 pp.
119. DIENES, LESLIE. *Locational Factors and Locational Developments in the Soviet Chemical Industry* 1969. 285 pp.
120. MIHELIC, DUSAN. *The Political Element in the Port Geography of Trieste* 1969. 104 pp.
121. BAUMANN, DUANE. *The Recreational Use of Domestic Water Supply Reservoirs: Perception and Choice* 1969. 125 pp.
122. LIND, AULIS O. *Coastal Landforms of Cat Island, Bahamas: A Study of Holocene Accretionary Topography and Sea-Level Change* 1969. 156 pp.
123. WHITNEY, JOSEPH. *China: Area, Administration and Nation Building* 1970. 198 pp.
124. EARICKSON, ROBERT. *The Spatial Behavior of Hospital Patients: A Behavioral Approach to Spatial Interaction in Metropolitan Chicago* 1970. 198 pp.
125. DAY, JOHN C. *Managing the Lower Rio Grande: An Experience in International River Development* 1970. 277 pp.
126. MAC IVER, IAN. *Urban Water Supply Alternatives: Perception and Choice in the Grand Basin, Ontario* 1970. 178 pp.
127. GOHEEN, PETER G. *Victorian Toronto, 1850 to 1900: Pattern and Process of Growth* 1970. 278 pp.
128. GOOD, CHARLES M. *Rural Markets and Trade in East Africa* 1970. 252 pp.
129. MEYER, DAVID R. *Spatial Variation of Black Urban Households* 1970. 127 pp.
130. GLADFELTER, BRUCE. *Meseta and Campiña Landforms in Central Spain: A Geomorphology of the Alto Henares Basin* 1971. 204 pp.
131. NEILS, ELAINE M. *Reservation to City: Indian Urbanization and Federal Relocation* 1971. 200 pp.
132. MOLINE, NORMAN T. *Mobility and the Small Town, 1900–1930* 1971. 169 pp.
133. SCHWIND, PAUL J. *Migration and Regional Development in the United States, 1950–1960* 1971. 170 pp.
134. PYLE, GERALD F. *Heart Disease, Cancer and Stroke in Chicago: A Geographical Analysis with Facilities Plans for 1980* 1971. 292 pp.
135. JOHNSON, JAMES F. *Renovated Waste Water: An Alternative Source of Municipal Water Supply in the U.S.* 1971. 155 pp.
136. BUTZER, KARL W. *Recent History of an Ethiopian Delta: The Omo River and the Level of Lake Rudolf* 1971. 184 pp.
137. HARRIS, CHAUNCY D. *Annotated World List of Selected Current Geographical Serials in English, French, and German* 3rd edition 1971. 77 pp.
138. HARRIS, CHAUNCY D., and FELLMANN, JEROME D. *International List of Geographical Serials* 2nd edition 1971. 267 pp.
139. MC MANIS, DOUGLAS R. *European Impressions of the New England Coast, 1497–1620* 1972. 147 pp.
140. COHEN, YEHOSHUA S. *Diffusion of an Innovation in an Urban System: The Spread of Planned Regional Shopping Centers in the United States, 1949–1968* 1972. 136 pp.

141. MITCHELL, NORA. *The Indian Hill-Station: Kodaikanal* 1972. 199 pp.
142. PLATT, RUTHERFORD H. *The Open Space Decision Process: Spatial Allocation of Costs and Benefits* 1972. 189 pp.
143. GOLANT, STEPHEN M. *The Residential Location and Spatial Behavior of the Elderly: A Canadian Example* 1972. 226 pp.
144. PANNELL, CLIFTON W. *T'ai-chung, T'ai-wan: Structure and Function* 1973. 200 pp.
145. LANKFORD, PHILIP M. *Regional Incomes in the United States, 1929–1967: Level, Distribution, Stability, and Growth* 1972. 137 pp.
146. FREEMAN, DONALD B. *International Trade, Migration, and Capital Flows: A Quantitative Analysis of Spatial Economic Interaction* 1973. 202 pp.
147. MYERS, SARAH K. *Language Shift Among Migrants to Lima, Peru* 1973. 204 pp.
148. JOHNSON, DOUGLAS L. *Jabal al-Akhdar, Cyrenaica: An Historical Geography of Settlement and Livelihood* 1973. 240 pp.
149. YEUNG, YUE-MAN. *National Development Policy and Urban Transformation in Singapore: A Study of Public Housing and the Marketing System* 1973. 204 pp.
150. HALL, FRED L. *Location Criteria for High Schools: Student Transportation and Racial Integration* 1973. 156 pp.
151. ROSENBERG, TERRY J. *Residence, Employment, and Mobility of Puerto Ricans in New York City* 1974. 230 pp.
152. MIKESELL, MARVIN W., editor. *Geographers Abroad: Essays on the Problems and Prospects of Research in Foreign Areas* 1973. 296 pp.
153. OSBORN, JAMES. *Area, Development Policy, and the Middle City in Malaysia* 1974. 273 pp.
154. WACHT, WALTER F. *The Domestic Air Transportation Network of the United States* 1974. 98 pp.
155. BERRY, BRIAN J. L., et al. *Land Use, Urban Form and Environmental Quality* 1974. 464 pp.
156. MITCHELL, JAMES K. *Community Response to Coastal Erosion: Individual and Collective Adjustments to Hazard on the Atlantic Shore* 1974. 209 pp.
157. COOK, GILLIAN P. *Spatial Dynamics of Business Growth in the Witwatersrand* 1975. 143 pp.
158. STARR, JOHN T., JR. *The Evolution of Unit Train Operations in the United States: 1960–1969—A Decade of Experience* 1976. 247 pp.
159. PYLE, GERALD F. *The Spatial Dynamics of Crime* 1974. 220 pp.
160. MEYER, JUDITH W. *Diffusion of an American Montessori Education* 1975. 109 pp.
161. SCHMID, JAMES A. *Urban Vegetation: A Review and Chicago Case Study* 1975. 280 pp.
162. LAMB, RICHARD. *Metropolitan Impacts on Rural America* 1975. 210 pp.
163. FEDOR, THOMAS. *Patterns of Urban Growth in the Russian Empire during the Nineteenth Century* 1975. 275 pp.
164. HARRIS, CHAUNCY D. *Guide to Geographical Bibliographies and Reference Works in Russian or on the Soviet Union* 1975. 496 pp.
165. JONES, DONALD W. *Migration and Urban Unemployment in Dualistic Economic Development* 1975. 186 pp.
166. BEDNARZ, ROBERT S. *The Effect of Air Pollution on Property Value* 1975. 118 pp.
167. HANNEMANN, MANFRED. *The Diffusion of the Reformation in Southwestern Germany, 1518-1534* 1975. 248 pp.
168. SUBLETT, MICHAEL D. *Farmers on the Road. Interfarm Migration and the Farming of Noncontiguous Lands in Three Midwestern Townships, 1939-1969* 1975. 228 pp.
169. STETZER, DONALD FOSTER. *Special Districts in Cook County: Toward a Geography of Local Government* 1975. 189 pp.
170. EARLE, CARVILLE V. *The Evolution of a Tidewater Settlement System: All Hallow's Parish, Maryland, 1650-1783* 1975. 249 pp.
171. SPODEK, HOWARD. *Urban-Rural Integration in Regional Development: A Case Study of Saurashtra, India—1800-1960* 1976. 156 pp.
172. COHEN, YEHOSHUA S. and BERRY, BRIAN J. L. *Spatial Components of Manufacturing Change* 1975. 272 pp.
173. HAYES, CHARLES R. *The Dispersed City: The Case of Piedmont, North Carolina* 1976. 169 pp.
174. CARGO, DOUGLAS B. *Solid Wastes: Factors Influencing Generation Rates* 1976.
175. GILLARD, QUENTIN. *Incomes and Accessibility. Metropolitan Labor Force Participation, Commuting, and Income Differentials in the United States, 1960-1970* 1976. 140 pp.
176. MORGAN, DAVID J. *Patterns of Population Distribution: A Residential Preference Model and Its Dynamic* 1976.
177. STOKES, HOUSTON H.; JONES, DONALD W. and NEUBURGER, HUGH M. *Unemployment and Adjustment in the Labor Market: A Comparison between the Regional and National Responses* 1975. 135 pp.
178. PICCAGLI, GIORGIO ANTONIO. *Racial Transition in Chicago Public Schools. An Examination of the Tipping Point Hypothesis, 1963-1971* 1976.
179. HARRIS, CHAUNCY D. *Bibliography of Geography. Part I. Introduction to General Aids* 1976. 288 pp.